AF360888

Air Pollution Modelling

Air Pollution Modelling

Local-, Regional-, and Global-Scale Application

Editor

Syuichi Itahashi

MDPI • Basel • Beijing • Wuhan • Barcelona • Belgrade • Manchester • Tokyo • Cluj • Tianjin

Editor
Syuichi Itahashi
Central Research Institute of
Electric Power Industry
Japan

Editorial Office
MDPI
St. Alban-Anlage 66
4052 Basel, Switzerland

This is a reprint of articles from the Special Issue published online in the open access journal *Atmosphere* (ISSN 2073-4433) (available at: https://www.mdpi.com/journal/atmosphere/special_issues/air_model).

For citation purposes, cite each article independently as indicated on the article page online and as indicated below:

LastName, A.A.; LastName, B.B.; LastName, C.C. Article Title. *Journal Name* **Year**, *Volume Number*, Page Range.

ISBN 978-3-0365-0644-9 (Hbk)
ISBN 978-3-0365-0645-6 (PDF)

Contents

About the Editor

Syuichi Itahashi is the research scientist at Central Research Institute of Electric Power Industry (CRIEPI), Japan. He has been engaged in the air quality modeling study over East Asia. He served as the Editorial Advisory Board member of the Asian Journal of Atmospheric Environment (AJAE) from 2016 to 2020.

Editorial

Air Pollution Modeling: Local, Regional, and Global-Scale Applications

Syuichi Itahashi

Central Research Institute of Electric Power Industry, Abiko, Chiba 270-1194, Japan; isyuichi@criepi.denken.or.jp

Academic Editor: Gabriele Curci
Received: 23 January 2021; Accepted: 28 January 2021; Published: 29 January 2021

The *Atmosphere* Special Issue entitled "Air Pollution Modeling: Local, Regional, and Global-Scale Applications" comprises nine original papers.

The problem of air pollution is inevitably accompanied with our human activities. Severe air pollution situations have been reported, especially in emerging countries, and fully satisfying air quality standards is still an underlying issue. Today, modeling research is a valuable approach to promote our understanding on the behavior of air pollutants and can be used for regulatory, policy, and environmental decision making. Such modeling applications are wide ranging with regard to horizontal grid resolution varying from a few km (local), to hundreds of km (regional), to thousands of km (global). To foster our current scientific knowledge on the potential and limitations of modeling, scientific research studies related to air pollution modeling—applied at the urban, regional, and global-scale—have been collected in this Special Issue.

The following three papers in this Special Issue conducted high-resolution modeling studies using computational fluid dynamics (CFD). Kim [1] applied a coupled CFD–chemistry model to examine the sensitivity of nitrate aerosols to vehicular emissions in urban streets. Sensitivity simulations clarified that nitrate concentrations do not show a clear relationship with NO_x emission rates, and show a proportional relationship with vehicular volatile organic compound (VOC) and NH_3 emissions in the street canyon. The results suggest that the control of vehicular VOC and NH_3 emissions might be a more effective way to reduce $PM_{2.5}$ problems, rather than the control of NO_x emissions, when vehicular emissions are dominant in winter. Gonzalez Olivardia et al. [2] presented the airflow patterns and reactive pollutant behavior over 24 h, in a realistic urban canyon in Osaka City, Japan, using a CFD model coupled with a chemical reaction model. The mesoscale model cannot capture the local phenomena because of the coarse grid resolution, and the CFD-coupled chemical reaction model surpassed the mesoscale models in describing the NO, NO_2, and O_3 transport process, representing pollutants concentrations more accurately within the street canyon. This work showed that the concentration of pollutants in the urban canyon is heavily reliant on roadside emissions and airflow patterns, which, in turn, are strongly affected by the heterogeneity of the urban layout. The CFD-coupled chemical reaction model better characterized the complex three-dimensional site and hour-dependent dispersion of contaminants within an urban canyon. Moissseeva and Stull [3] designed a large-eddy simulation of a real-life prescribed burn using a coupled semi-emperical fire–atmosphere model. Large eddy simulation (LES) is a method that uses CFD at a very fine spatial and temporal resolution to simulate wide ranging scales of atmospheric motions, down to the size of large turbulent eddies. The Weather Research and Forecasting (WRF) model, combined with a semi-empirical fire-spread algorithm (WRF-SFIRE), allows for the two-way coupling between a LES and a fire behavior model. WRF-SFIRE was applied to simulate wildfire smoke plume dynamics, and the vertical rise and distribution of emissions on a local scale. The results suggest that the rise and dispersion of fire emissions are reasonably well captured by the model; subject to accurate surface thermal forcing and relatively steady atmospheric conditions.

For local, regional, and global-scale modeling, the Eulerian model serves as an important approach. There are four papers within this Special Issue related to the topic. Takami et al. [4] focused on biomass burning, which is a major source of atmospheric particle pollution over Indochina during the dry season. Indochina has convoluted meteorological scales, and regional meteorological conditions dominate the transport patterns of pollutants. Based on the Community Multiscale Air Quality (CMAQ) model, the impacts of biomass burning emission inventories and meteorology on simulated PM_{10} over Indochina in 2014 were examined by adopting different emission inventories and an atmospheric re-analysis dataset. The results clarified that the biomass burning emission inventories impact on PM_{10} simulation to a greater degree than atmospheric re-analyses in area highly polluted by biomass burning in Indochina in 2014. Yamaji et al. [5] present a Japanese study on reference air quality modeling (J-STREAM), conducted using 32 model settings for representing $PM_{2.5}$ concentration. A Comprehensive Air Quality Model (CMAQ) with eXtensions (CAMx) and Weather Research and Forecasting-Chemistry (WRF-Chem) were included in the project. Good performances were found in dominant components of $PM_{2.5}$, such as sulfates (SO_4^{2-}) and ammonium (NH_4^+). The other simulated $PM_{2.5}$ components, i.e., nitrates (NO_3^-), elemental carbon (EC), and organic carbon (OC), often showed clear deviations from the observations. This inter-comparison model suggested that these deviations may be owing to a need for further improvements both in the emission inventories and additional formation pathways in chemical transport models, while meteorological conditions also require improvement to simulate elevated atmospheric pollutants. Itahashi et al. [6] presented the possible pathways to solve the subject of model underestimation of SO_4^{2-} during winter found in J-STREAM. A winter haze period in December 2016 was examined by involving aqueous-oxidations and gas-phase oxidation by three stabilized Criegee intermediates into the CMAQ model. Updating the catalyzed aqueous phase metal and NO_2 oxidation pathways led to increased contributions from other sources, and the additional gas phase oxidation by stabilized Criegee intermediates provided a link between fugitive VOC emissions via changes in O_3. Zong et al. [7] performed the WRF-Chem simulation for winter visibility in the Jiangsu Province in China. The result showed that WRF-Chem has a good capability to simulate visibility and the related local meteorological elements and air pollutants in Jiangsu in the winters of 2013–2017. For visibility inversion, the study further adopted the neural network algorithm. Meteorological elements, including wind speed, humidity and temperature, were introduced to improve the performance of WRF-Chem relative to the visibility inversion scheme. The underestimation of the visibility was especially remarkable, but this issue was essentially solved using the neural network algorithm.

Other models such as the dispersion model and the Lagrangian model are also applied and presented in this Special Issue. Prato and Huertas [8] assessed the agricultural burning influence-area using the air dispersion model of AERMOD. Agricultural burning, including short-term climate change forcing pollutants such as black carbon is still a common practice around the world. The legal requirements to commence regulatory actions to control them relate to the identification of the area of influence. However, this task is challenging from the experimental and modeling point of view, since it is a short-term event and the source area of the pollutants moves. Different sizes and geometries of burning areas located on flat terrains, and with several crops burning under worst-case scenarios of meteorological conditions were considered in this study. The influence area was determined as the largest area where the short-term concentrations of pollutants (1 h or one day) exceed the local air quality standards. It was found that this area forms a band around the burning area, the size of which increases with the burning rate but not with its size. Haszpra [9] introduced a Lagrangian model called the Real Particle Lagrangian Trajectory model—Chaos version (RePLaT-Chaos), which specifically aims to demonstrate the chaotic behavior of pollutants, i.e., the advection dynamics of pollutants shows the typical characteristics, such as sensitivity to the initial conditions, irregular motion, and complicated but well-organized (fractal) structures. This study presents possible applications of a RePLaT-Chaos by means of which the characteristics of the long-range atmospheric spreading of volcanic ash clouds and other pollutants can be investigated in an easy and interactive way. This application is also a suitable tool for studying the chaotic features of advection, and determines two quantities which describe the chaotic nature of the advection processes: the stretching rate quantifies the strength of the exponential stretching of pollutant clouds, and the escape rate characterizes the rate of the rapidity by which the settling particles of a pollutant cloud leave the atmosphere.

The goal of this Special Issue is to present a research with a wide perspective, involving multi-scale modeling research studies with different types of modeling applications, and the nine papers in this Special Issue achieve this goal. The research presented in these nine papers demonstrate the usefulness of modeling applications.

Funding: This research received no external funding.

Institutional Review Board Statement: Not applicable.

Informed Consent Statement: Not applicable.

Acknowledgments: The editor would like to thank the authors from the Republic of Korea, Colombia, Mexico, Japan, Canada, Hungary, and China for their contributions to this Special Issue, and the reviewers for their constructive and helpful comments to improve the manuscripts. The editor is grateful to Xue Liang for her kind support in processing and publishing this Special Issue.

Conflicts of Interest: The authors declare no conflict of interest.

References

1. Kim, M. Sensitivity of Nitrate Aerosol Production to Vehicular Emissions in an Urban Street. *Atmosphere* **2019**, *10*, 212. [CrossRef]
2. Prato, D.; Huertas, J. Determination of the Area Affected by Agricultural Burning. *Atmosphere* **2019**, *10*, 312. [CrossRef]
3. Gonzalez Olivardia, F.; Zhang, Q.; Matsuo, T.; Shimadera, H.; Kondo, A. Analysis of Pollutant Dispersion in a Realistic Urban Street Canyon Using Coupled CFD and Chemical Reaction Modeling. *Atmosphere* **2019**, *10*, 479. [CrossRef]
4. Itahashi, S.; Yamaji, K.; Chatani, S.; Hayami, H. Differences in Model Performance and Source Sensitivities for Sulfate Aerosol Resulting from Updates of the Aqueous- and Gas-Phase Oxidation Pathways for a Winter Pollution Episode in Tokyo, Japan. *Atmosphere* **2019**, *10*, 544. [CrossRef]
5. Moisseeva, N.; Stull, R. Capturing Plume Rise and Dispersion with a Coupled Large-Eddy Simulation: Case Study of a Prescribed Burn. *Atmosphere* **2019**, *10*, 579. [CrossRef]
6. Haszpra, T. RePLaT-Chaos: A Simple Educational Application to Discover the Chaotic Nature of Atmospheric Advection. *Atmosphere* **2020**, *11*, 29. [CrossRef]
7. Takami, K.; Shimadera, H.; Uranishi, K.; Kondo, A. Impacts of Biomass Burning Emission Inventories and Atmospheric Reanalyses on Simulated PM10 over Indochina. *Atmosphere* **2020**, *11*, 160. [CrossRef]
8. Yamaji, K.; Chatani, S.; Itahashi, S.; Saito, M.; Takigawa, M.; Morikawa, T.; Kanda, I.; Miya, Y.; Komatsu, H.; Sakurai, T.; et al. Model Inter-Comparison for PM2.5 Components over urban Areas in Japan in the J-STREAM Framework. *Atmosphere* **2020**, *11*, 222. [CrossRef]
9. Zong, P.; Zhu, Y.; Wang, H.; Liu, D. WRF-Chem Simulation of Winter Visibility in Jiangsu, China, and the Application of a Neural Network Algorithm. *Atmosphere* **2020**, *11*, 520. [CrossRef]

Atmosphere **2021**, *12*, 178

Article

WRF-Chem Simulation of Winter Visibility in Jiangsu, China, and the Application of a Neural Network Algorithm

Peishu Zong [1,2,3,*], Yali Zhu [1,4,5,*], Huijun Wang [4,5] and Duanyang Liu [3,6,7]

[1] Nansen-Zhu International Research Centre, Institute of Atmospheric Physics, Chinese Academy of Sciences, Beijing 100029, China
[2] University of Chinese Academy of Sciences, Beijing 100049, China
[3] Key Laboratory of Transportation Meteorology, China Meteorological Administration, Nanjing 210008, China; liuduanyang2001@126.com
[4] Collaborative Innovation Center on Forecast and Evaluation of Meteorological Disasters, Nanjing University of Information Science & Technology, Nanjing 210044, China; wanghj@mail.iap.ac.cn
[5] Key Laboratory of Meteorological Disaster, Nanjing University of Information Science and Technology, Nanjing 210044, China
[6] Jiangsu Institute of Meteorological Sciences, Nanjing 210008, China
[7] Nanjing Joint Institute for Atmospheric Sciences, Nanjing 210008, China
[*] Correspondence: zongps@mail.iap.ac.cn (P.Z.); zhuyl@mail.iap.ac.cn (Y.Z.)

Received: 18 March 2020; Accepted: 11 May 2020; Published: 18 May 2020

Abstract: In this paper, the winter visibility in Jiangsu Province is simulated by WRF-Chem (Weather Research and Forecasting (WRF) model coupled with Chemistry) with high spatiotemporal resolutions. Simulation results show that WRF-Chem has good capability to simulate the visibility and related local meteorological elements and air pollutants in Jiangsu in the winters of 2013–2017. For visibility inversion, this study adopts the neural network algorithm. Meteorological elements, including wind speed, humidity and temperature, are introduced to improve the performance of WRF-Chem relative to the visibility inversion scheme, which is based on the Interagency Monitoring of Protected Visual Environments (IMPROVE) extinction coefficient algorithm. The neural network offers a noticeable improvement relative to the inversion scheme of the IMPROVE visibility extinction coefficient, substantially improving the underestimation of winter visibility in Jiangsu Province. For instance, the correlation coefficient increased from 0.17 to 0.42, and root mean square error decreased from 2.62 to 1.76. The visibility inversion results under different humidity and wind speed levels show that the underestimation of the visibility using the IMPROVE scheme is especially remarkable. However, the underestimation issue is essentially solved using the neural network algorithm. This study serves as a basis for further predicting winter haze events in Jiangsu Province using WRF-Chem and deep-learning methods.

Keywords: WRF-Chem; visibility; eastern China; neural network algorithm; IMPROVE

---

1. Introduction

Daytime visibility is defined as the maximum horizontal distance at which an object can be seen and recognized against the sky background under current weather conditions. In meteorology, low atmospheric visibility generally refers to visibility impairment caused by weather processes, such as fog, haze, rain, snow, and sandstorms. Due to the rapid development of urbanization and the sharp increase in pollutant emissions, high concentrations of aerosols can also lead to low visibility, which has become one of the most important atmospheric environmental problems in many regions worldwide. Air pollution has adverse effects on human physiological function and increases the risk of death [1,2]. It also has considerable potential impacts on the tourism industry [3] and ecological environment [4]. Therefore, accurate simulation and prediction of air pollution are necessary for public health and sustainable development.

Studies show that the formation and persistence of low-visibility weather are closely related to the chemical processes of aerosols. Haze days are always accompanied by high concentrations of $PM_{2.5}$, PM_{10}, black carbon, sulfate, nitrate, organic matter, and dust [5]. Low-visibility weather is also affected by local meteorological conditions, such as temperature inversion, higher humidity, and lower wind speed [6]. Moreover, the relative contribution of meteorological factors exceeds 50% of the variance of low-visibility weather [5]. Recent studies [7,8] indicate that the reduction of sea ice in the Arctic in autumn will lead to anomalies in the atmospheric circulation over the Eurasian continent, the weakening of cyclone activity in the northern part of the country and stabilization of atmospheric stratification, thus increasing haze pollution in the eastern part of the country in winter.

Recently, the ability of the current global dynamical prediction system to predict haze days in regional areas is far from meeting the actual needs of disaster prevention and reduction. Moreover, it is difficult to rapidly improve its prediction performance in the short term [9]. In this respect, downscaling is a relatively efficient and feasible method. The dynamical downscaling method uses regional climate models with higher resolution and more detailed physical process parameterization schemes by nesting global climate models to improve the simulation performance of regional climate, especially climate in complex underlying surfaces [10]. A number of studies have shown that regional climate models perform better than global models in simulating the spatial distribution of climate factors [11,12]. At present, the WRF-Chem (Weather Research and Forecasting (WRF) model coupled with Chemistry) model has been widely used to study air pollution [13,14]. Most studies of its application focus on the concentration and variation of the atmospheric pollutants in a certain pollution episode. However, few studies have been performed to research short-term climate predictions to inverse and diagnose atmospheric visibility. In spite of the excellent performance of the WRF model in the field of short-term climate prediction, its ability to simulate the meteorological elements, air quality and inverted regional visibility must be further examined.

Located along the eastern coast of mainland China, Jiangsu Province is an important economic region in China with rapid economic growth and fast worsening air quality in recent decades. Liu et al. (2017) show that the low visibility events in Jiangsu Province occur frequently in winter and are often haze events [9]. In addition, the regional continuous haze days show increasing trends in recent years. As urban areas expand each year, industrial emissions, coal consumption, and car ownership increased accordingly, resulting in regional temperature increases and relative humidity decreases that form the urban heat island and dry island effects. Hence, meteorological conditions became more favorable for haze formation and maintenance and thus favored more haze days, which caused significant increases in continuous, regional, and regional continuous haze days. Song et al. (2017) used the WRF-Chem model to examine the relationships between dust emissions and meteorological variables (wind speed, humidity, and temperature) over East Asia during the period of 1980–2015. Result shows that these meteorological variables substantially contribute to the haze days [15]. The northerly monsoon is the prevailing wind direction in winter in Jiangsu Province, and higher wind speed shows a strong correlation with higher visibility [16].

Jiangsu Province is the core area of the Yangtze River Delta region, which is one of the most air-polluted regions in China. Similar to the conditions in eastern China, continuous air polluting processes there were jointly affected by pollutant emissions, which are mainly produced by industrial development and transport pollution, and the weather conditions that favor the accumulation of pollutants as well. Steady meteorological conditions were important external factors affecting air pollution. Winter had the most stable meteorological conditions, thus most regional haze days appeared in winter, accompanied by lower visibility [17–19].

Considering the WRF-Chem model does not output the visibility values directly, many methods are developed to inverse visibility by the WRF-Chem output. The Interagency Monitoring of Protected

Visual Environment (IMPROVE) monitoring project in the United States is the most widely used. Recently, with the development of deep learning technology, some deep learning algorithms such as neural network and multi-criteria decision analysis are also used to analyze the air quality [20,21]. In this study, we systematically evaluate the ability of WRF-Chem to reproduce winter visibility in Jiangsu Province and introduce local meteorological elements (wind speed, humidity, and temperature) and a neural network scheme to further improve the simulation results. This study will provide a scientific basis for further developing a winter visibility forecast in Jiangsu Province.

2. Experimental Design, Data and Method

2.1. WRF-Chem Numerical Experiment Design

The WRF-Chem model was developed by NOAA (National Oceanic and Atmospheric Administration) Forecast System Laboratory (FSL) in the United States. The model is a new-generation regional air quality model that is fully coupled with the meteorological model (WRF) and chemical model (Chem) online [22]. This model has the full function of the WRF model and can be used to predict and simulate temperature, wind, cloud, and rain processes as well as other physical quantities. In addition, it can couple the weather forecast/diffusion model to simulate emission and transport components as well as the weather/diffusion/air quality model with integrated chemical species to simulate the interaction among ozone, ultraviolet radiation, particulate matter, etc.

The WRF-Chem model version 3.9.1 used in this paper was released in 2017, representing the latest version. The air quality in Jiangsu Province in the winters of 2013–2017 is simulated. The integration time is from 00:00 UTC on November 16th each year to 00:00 UTC on March 1st of the following year, and the first 15 days are spin-up time. The initial meteorological field uses NCEP FNL (National Centers for Environmental Prediction Final) $1^\circ \times 1^\circ$ data with an interval of 6 h. The model outputs once every hour with a spatial grid of 18 km and a grid number of 100×100 (Figure 1). The physical schemes include the Kain–Frisch convective scheme [23], Lin microphysics scheme [24], RRTMG-K radiation scheme [25], YUS boundary layer meteorological scheme [26], and Noah land surface scheme [27]. With regard to the meteorological chemical mechanism, the CBM-Z carbon bond mechanism, which includes 67 chemical reaction species and 164 chemical reactions, is adopted. Moreover, the MOSAIC mechanism is adopted as a chemical mechanism of aerosols, which includes sulfate (SULF = SO_4^{2-} + HSO^{4-}), methyl sulfuric acid (CH_3SO_3), nitrate (NO_3^{2+}), chloride (CL^-), carbonate, ammonium salt (NH_4^+), sodium salt, calcium salt, black carbon, organic carbon, and other aerosols [28]. The direct effects of aerosols are calculated based on Fast et al. (2006) and the Mie scattering theory [29]. The estimation of indirect effects of aerosols includes the effect of clouds on shortwave radiation, the calculation of aerosol activation/suspension, and the calculation of cloud droplet concentration is based on the number of activated aerosols [30,31].

The Multiresolution Emission Inventory for China (MEIC) is a set of anthropogenic emission inventory models of atmospheric pollutants and greenhouse gases in China based on the cloud computing platform. The MEIC is currently the most complete and accurate source of emissions in China. With a 0.25-degree resolution for the base years 2008 and 2010, the MEIC covers 10 major atmospheric pollutants and greenhouse gases, such as SO_2, NO_x, CO, NMVOC, NH_3, CO_2, $PM_{2.5}$, PM_{10}, BC and OC, and more than 700 anthropogenic emission sources. The MEIC 2010 V1.2 monthly gridded emissions with the spatial resolution of $0.25^\circ \times 0.25^\circ$ is used in this paper [31–35].

2.2. Visibility Inversion Scheme

In recent years, many studies have been performed to research the impact of aerosols on visibility in China and abroad. Among them, the Interagency Monitoring of Protected Visual Environment (IMPROVE) monitoring project in the United States is the longest project [36,37]. Since 1985, the project has made long-term observations and research on important optical parameters related to atmospheric visibility and the chemical composition of aerosols. Of numerous visibility inversion schemes, the most commonly used method is the atmospheric extinction coefficient estimation model. The atmospheric extinction coefficient refers to the total optical attenuation in a unit distance. The attenuation is caused by the scattering and absorption of gas and aerosols between the source and the receiver. Irrespective of the interference of special weather processes (such as rain, snow, fog, and dust storms, etc.) and the pollutant transport, the main factors affecting the extinction coefficient are summarized. These influencing factors related to aerosols include mass concentration of aerosols, particle size spectrum distribution, chemical composition, hygroscopicity, black carbon and its mixed state, and the shape of particles. The extinction calculation method is proposed by the IMPROVE project. The calculation idea is to reconstruct aerosol mass concentration and environmental extinction by measured aerosol species, including sulfate, nitrate, ammonia, organic matter, black carbon, coarse particulate matter ($PM_{2.5-10}$), and soil-derived aerosols. Moreover, the method is the basis for assessing the implementation of the regional haze rule (1999). Thus, we also adopted this method to calculate the visibility in Jiangsu Province through the inversion of aerosol concentration simulated by the air quality model WRF-Chem.

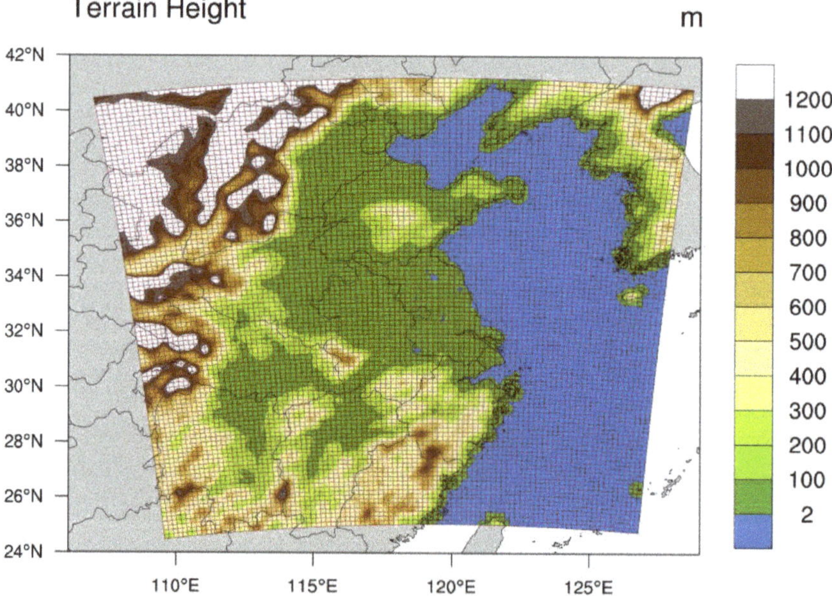

Figure 1. Simulation domain in WRF-Chem (the grid interval is 18 km).

Based on the modified empirical formula of extinction coefficient from IMPROVE of the United States in 2012 [38], the formula of atmospheric visibility inversion is established.

$$v = K / B_{ext}$$

where K is a constant with a value of 3.912, v is atmospheric visibility (km), and Bext (km^{-1}) is the extinction coefficient. The extinction coefficient is calculated as follows:

$$
\begin{aligned}
B_{ext} = B_{sg} + 2.2 \quad &\times f_s(RH) \times S(sulfate) + 4.8 \times f_L(RH) \times L(sulfate) + 2.4 \times f_s(RH) \\
&\times S(nitrate) + 5.1 \times f_L(RH) \times L(nitrate) + 2.8 \times S(OM) + 6.1 \\
&\times L(OM) + 10 \times [EC] + [FS] + 0.6 \times [CM] + 0.33 \times [NO_2]
\end{aligned}
$$

where B_{sg} is the Rayleigh scattering extinction coefficient (Mm^{-1}); $f_s(RH)$ and $f_L(RH)$ are the moisture absorption growth coefficients of coarse and fine particles, respectively, and they are functions of relative humidity (RH) [37]; $L(X)$ and $S(X)$ represent the mass concentration of coarse and fine aerosol particles (unit: $\mu g/m^3$), respectively, in which X denotes sulfate, nitrate and organic matter (OM); [EC], [FS] and [CM] are elemental carbon concentration, fine soil dust aerosol concentration ($PM_{2.5}$) and coarse particle concentration (PM_{10}) (unit: $\mu g/m^3$), respectively; and [NO_2] is the volume fraction (10^{-9}) of NO_2.

Simulated by WRF-Chem, the concentration of air pollutants, i.e., $\rho(SO_4^{2-})$, $\rho(NO_3^-)$, $\rho(NH_4^+)$, $\rho(OM)$, $\rho(BC)$, $\rho(PM_{10})$, $\rho(PM_{2.5})$ and $\rho(NO_2)$ are obtained. Using these factors and RH, the atmospheric extinction coefficient is calculated, and the visibility is subsequently obtained.

According to two calculation reports of the World Meteorological Organization (WMO) in 1990 and 2008, the conversion formulas between manual and automated visibility observation are as follows.

$$\frac{VIS_{instrumental}}{VIS_{observer}} = \frac{(1/k)*\ln(1/0.05)}{(1/k)*\ln(1/0.02)} \approx 0.766$$

In the equation, $VIS_{instrumental}$ is visibility obtained from the automated observation, $VIS_{observer}$ is visibility obtained from the manual observation, and k is the extinction coefficient.

Since January 1st 2014, the visibility observations were all performed by automatic observers in Jiangsu Province. Therefore, the visibility data before January 1st 2014 are converted into automatic observation visibility.

2.3. Visibility Correction Scheme

As the extinction coefficient formula of IMPROVE does not consider weather processes and external pollutant transport, the visibility inversion might be impacted to some extent. Therefore, in this paper, a method of visibility inversion is introduced to increase the weight of meteorological factors.

As the impact of meteorological factors on visibility is complex and nonlinear, a deep learning algorithm is adopted for the inversion scheme of visibility. Deep learning is a subfield of machine learning. It is a new approach of learning representation from data that emphasizes learning from continuous layers that correspond to meaningful representation. The higher layer level contains richer information. In deep learning, these layers are almost always obtained by a model called a neural network. The structure of the neural network is stacked layer by layer. The neural network is derived from neurobiology. However, a deep learning model is a mathematical framework for learning representation from data in contrast to a brain model. In this paper, a simple but typical neural network algorithm is used to build up the visibility model based on the relationship between the meteorological/chemical factors and the visibility. Thus, this model can overcome the limitations of the traditional multivariate linear regression method in statistical modeling and improves visibility inversion products.

In the extinction coefficient formula in Section 2.2, the visibility inversion scheme based on the extinction coefficient formula takes into account chemical factors, such as sulfate, nitrate, ammonia, OM, black carbon, PM ($PM_{2.5-10}$) and aerosol factors, whereas meteorological factors are not included. However, in recent years, many studies [39,40] have shown that surface meteorological factors, such as temperature, humidity, and wind, have an important impact on regional low visibility episodes. Therefore, it is necessary to introduce these factors into the model of visibility inversion.

The surface meteorological elements, i.e., 2-m humidity, 2-m air temperature, 10-m meridional and zonal wind speed, are obtained from WRF-Chem simulation. Furthermore, the air pollutants which are in the IMPROVE equation, including sulfate, nitrate, ammonia, OM, black carbon, PM ($PM_{2.5-10}$) and aerosol factors, are also involved in the neural network model. Along with the modulated visibility based on the scheme in 2.2, the observed visibility is modeled. The training period is 90 days in the winter of 2013, and the daily visibility in winters of 2014–2017 is evaluated.

The multilayer neural network is a deep learning method, and its training mainly focuses on the four factors as follows [41]:

(1) Layer: it is the core component of neural network. It is a data processing module that extracts representation from input data. Thus, simple layers are linked to realize progressive data distillation.

(2) Input data and corresponding targets: in this paper, the input training data are 2-m humidity, 2-m air temperature, 10-m meridional and zonal wind speed, sulfate, nitrate, ammonia, OM, black carbon, PM ($PM_{2.5-10}$), and aerosol factors in the winter of 2013 simulated by WRF-Chem. The target data are the visibility diurnal series in the Yangtze River Delta in winter of 2013.

(3) Loss function: it is used to measure the performance of the neural network on training data to ensure the network progresses correctly. In this paper, the mean square error between the prediction and observation data are used as the loss function.

(4) Optimizer: it is a mechanism for updating the network based on training data and loss function. The smaller the loss function value, the better the performance of the neural network model.

The relationship among the four factors is shown in Figure 2. Multiple layers are linked together, forming a network, and the input data are mapped into predicted values by the network. Then, the loss function compares these predicted values with targets and obtains the loss values, measuring the matching degree between predicted values and targets. Optimizer updates the weight of network by the loss value. The first order optimization algorithm is Gradient Descent.

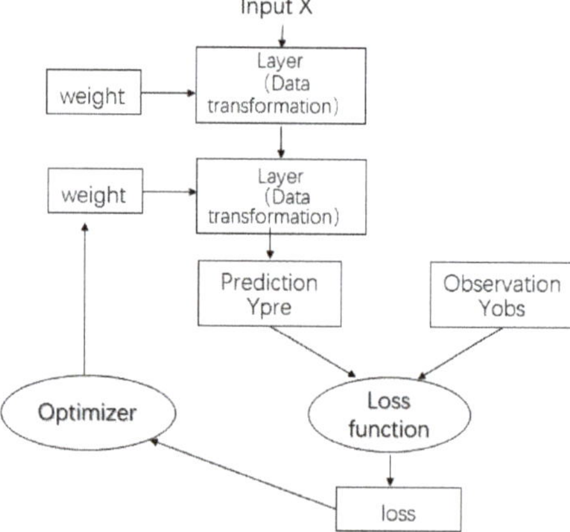

Figure 2. Relationship among network, layer, loss function, and optimizer.

This paper uses the neural network algorithm to investigate the relationship between the meteorological and chemical factors and the visibility from the training data set. The neural network algorithm uses meteorological and chemical factors over the Jiangsu province to automatically extract features that can be used to invert the visibility. A one-way propagation multi-layer feedforward network is used in this paper. The meteorological and chemical factors can be treated as neurons of the input layer. Moreover, each of the three hidden layers has 64 units. The input signal passes through each hidden layer in turn to the output layer node.

2.4. Error Analysis Method

In general, four error analysis methods are available, including correlation coefficient (R), root mean square error (RMSE), mean fractional error (MFE), and mean fractional bias (MFB) [42]. In numerical simulation, a model performance goal is met when both MFE and MFB are less than or equal to +50% and ±30%, respectively. This standard is also used in the verification of model products in our study.

$$R = \frac{\text{Cov}(S - O)}{\sqrt{\text{Var}(S)\,\text{Var}(O)}}$$

$$\text{RMSE} = \left[\frac{1}{n}\sum_{i=1}^{n}(S(i) - O(i))^2\right]^{\frac{1}{2}}$$

$$\text{MFE} = \frac{1}{n}\sum_{i=1}^{n}\left|\frac{S(i) - O(i)}{O(i) + S(i)/2}\right|$$

$$\text{MFB} = \frac{1}{n}\sum_{i=1}^{n}\left(\frac{S(i) - O(i)}{O(i) + S(i)/2}\right)$$

In the above equations, S and O represent the simulated and observed sequence, respectively. The number of samples is n, and sample i is recorded as S (i) and O (i), respectively. Cov (S, O) is the covariance of S and O. Var (S) and Var (O) are the variances of S and O, respectively.

2.5. Observation Data

The environmental monitoring data are hourly monitoring data of atmospheric chemical composition obtained from 72 environmental monitoring stations in Jiangsu Province from 2013 to 2017. The chemical components include SO_2, NO_2, PM_{10}, $PM_{2.5}$, CO, and O_3. The location distribution of the 72 stations is marked by red dots in Figure 3. As noted in Figure 2, rather than being evenly distributed throughout the area of Jiangsu Province, these monitoring stations are centered on cities. Therefore, when analyzing a city, the results of all the monitoring station in this city are equally weighted and averaged as the monitoring values of atmospheric chemical composition of the city.

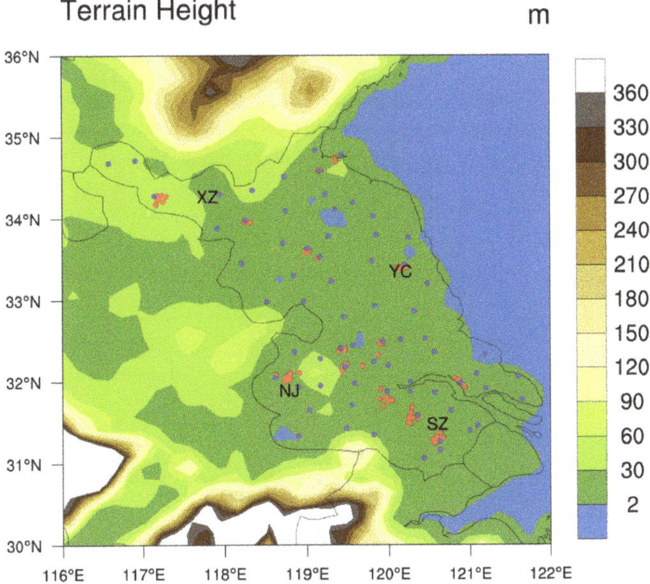

Figure 3. Locations of 72 environmental monitoring stations (red dot) and 70 meteorological observation stations (blue dot) in Jiangsu Province.

Meteorological observation data adopt the 3-hourly surface observation data from 70 stations, including 3 national base stations, 21 basic stations, and 46 general stations in Jiangsu Province. The data contain surface air temperature measured at 2 m above ground, relative humidity, wind direction, weather phenomena, visibility, precipitation, and other meteorological elements. As noted in Figure 2, the locations of 70 stations marked by blue dots are distributed evenly in Jiangsu Province.

3. Simulation Result Analysis

3.1. Simulation of Meteorological Elements in Jiangsu Province

The model results are interpolated to the locations of 70 meteorological stations in Jiangsu Province. Further, the simulated regional average value is compared with the observed average value of 70 stations to test the ability of the WRF-Chem model to simulate the winter meteorological elements in Jiangsu Province from 2013 to 2017.

The observed and simulated spatial distribution of 2013–2017 winter mean visibility in Jiangsu province is shown in Figure 4. The winter mean visibility is lower than 8 km over most of the area, which can be well reproduced by WRF-Chem. However, the simulation somewhat overestimates and underestimates the visibility in the northern and southern part, respectively, which means the haze events could be underestimated and overestimated correspondingly.

Figure 5 shows the observed and simulated data by the WRF-Chem model in winters of 2013–2017 in Jiangsu Province. The data include the 2-m air temperature (T_2), 2-m relative humidity (RH_2), 10-m wind speed (WS_{10}) and the time series of visibility (VIS). The 3-hourly observation data are rendered in scattered red dots. The WRF-Chem model outputs once an hour as marked by a black curve. The value at the upper right corner is the correlation coefficient (R) of the observed and simulated 3-hourly data. As the observation's time step is 3-hour, the length of the time series used for statistical calculation is 720. To evaluate the simulation quantitatively, the statistics of meteorological variables obtained from observation and simulation are also calculated, as shown in Table 1. As presented in Figure 5 and Table 1, the most effective simulation is the temporal variation of T_2 with an R reaching 0.95, followed by the simulation of the temporal variation of RH_2 with an R of approximately 0.9. In addition, the mean relative error of visibility in Table 1 is minus 32% and is close to the standard of 30%. Additionally, the mean relative deviation is less than 50%. Hence, the inversion results are generally useful. Moreover, the inverted visibility obtained from output products from WRF-Chem model is also highly correlated with the actual observation with an R of 0.77. The black dotted line in Figure 5d represents the visibility threshold of haze events (7.5 km), and the number of haze days inverted by WRF-Chem overestimates the observed data obviously.

Table 1. Correlation coefficient (R), root mean square error (RMSE), mean fractional error (MFE), and mean fractional bias (MFB) of the multi-winter (2013–2017) mean 3-hourly T_2, RH_2, WS_{10}, and VIS simulated by WRF-Chem model compared to the observation in Jiangsu Province.

Weather Element	R	RMSE	MFE	MFB
$T_2(°C)$	0.95	0.97	0.04	0.11
$RH_2(\%)$	0.90	5.95	−0.02	0.04
$WS_{10}(m/s)$	0.43	0.75	−0.02	0.19
Visibility(km)	0.76	3.70	−0.32	0.33

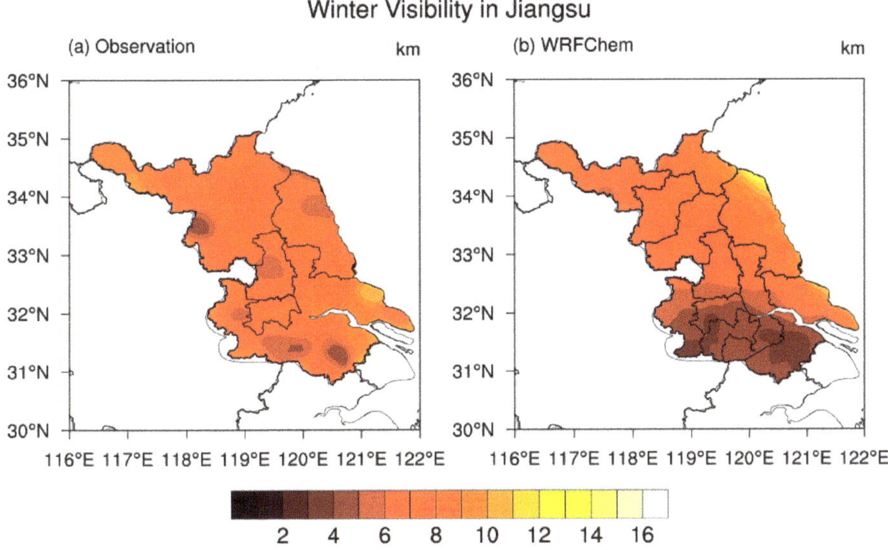

Figure 4. Mean visibility in winters of 2013–2017 in Jiangsu Province based on (**a**) observation and (**b**) WRF-Chem model output products.

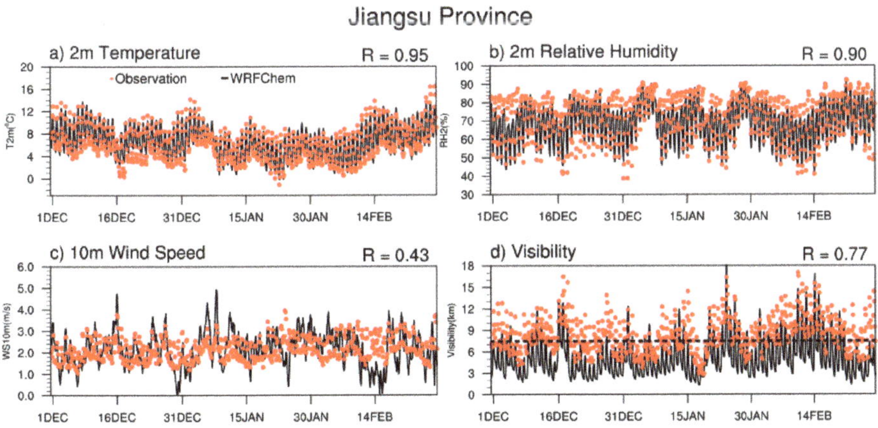

Figure 5. The temporal variation of multi-winter (2013–2017) mean 3-hourly (**a**) 2m temperature, (**b**) 2m relative humidity, (**c**) 10m wind speed, (**d**) visibility in Jiangsu province in observation (red spots) and WRF-Chem simulation (black curves).

To test in detail the ability of the WRF-Chem model to simulate urban meteorological elements and the inverted visibility, four representative cities in Jiangsu Province were selected, namely, Xuzhou City (XZ), Yancheng City (YC), Nanjing City (NJ), and Suzhou City (SZ). The locations of these four cities, which are shown in Figure 3, are as follows: Xuzhou City is located in the extreme northwest of Jiangsu Province close to Shandong Province; Yancheng City is located on the east coast of Jiangsu Province and has the longest coastline in Jiangsu Province; Nanjing is located in the southwest of Jiangsu Province, bordering Anhui Province; and Suzhou is located in the south of Jiangsu Province near Shanghai. Figures 6 and 7 show the mean values of T_2, RH_2, and WS_{10} obtained from observation and WRF-Chem model inversion in the winters of 2013–2017 in these four cities as well as the time series of

VIS from observation and inversion based on WRF-Chem products. Meanwhile, the statistics of each meteorological variable obtained from observations and simulations are also calculated, as shown in Table 2. Similar to the simulation results in Jiangsu Province, the simulation of temporal variation of T_2 in each city is the most effective, with an R of approximately 0.90, followed by the temporal variation of RH_2, with an R of approximately 0.8. Moreover, the average relative errors of all variables in Table 2 are within the range of minus 30% and plus 30%, and the average relative deviation is less than 50%, which meet the standards of accurate simulation. The inverted visibility obtained from output products based on the WRF-Chem model is also highly correlated with the actual data, with an R of approximately 0.5. The dotted line in the visibility curve figure is the visibility threshold of haze (7.5 km). The visibility curve and scattering points of XZ for most of the wintertime are below the dotted line. This could be attributed to that XZ is located in the northwest of Jiangsu Province with a prevalent northwest wind in winter. Thus, air pollutants are continuously transported from North China and Central China, resulting in frequent low-visibility weather in XZ in winter. The visibility of YC is obviously higher than that of XZ. The visibility of YC inverted by WRF-Chem can reach a maximum of 77 km. This is mainly because YC is located on the eastern coast of Jiangsu Province, and the strong local sea–land wind helps to remove pollutants. On the whole, the ground meteorological elements simulated by WRF-Chem and the values and temporal variation of visibility obtained from inversion are in good agreement with the observations.

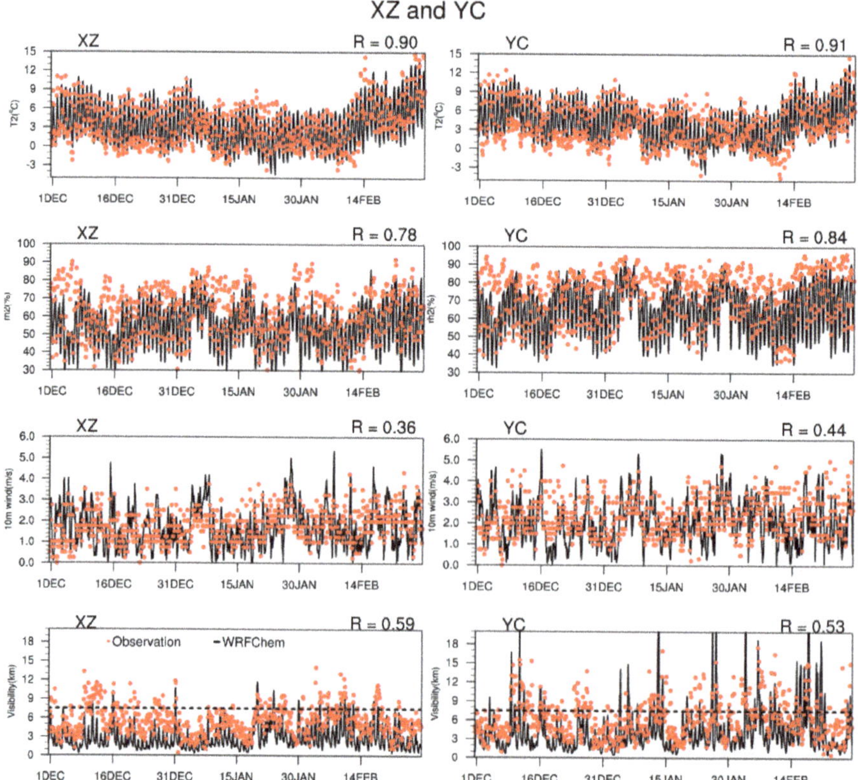

Figure 6. The temporal variation of multi-winter (2013–2017) mean 3-hourly T_2, RH_2, WS_{10}, and VIS from top to bottom, obtained from observation (red dots) and WRF-Chem simulation (black curves) in the city of Xuzhou City (XZ) (left column) and Yancheng City (YC) (right column).

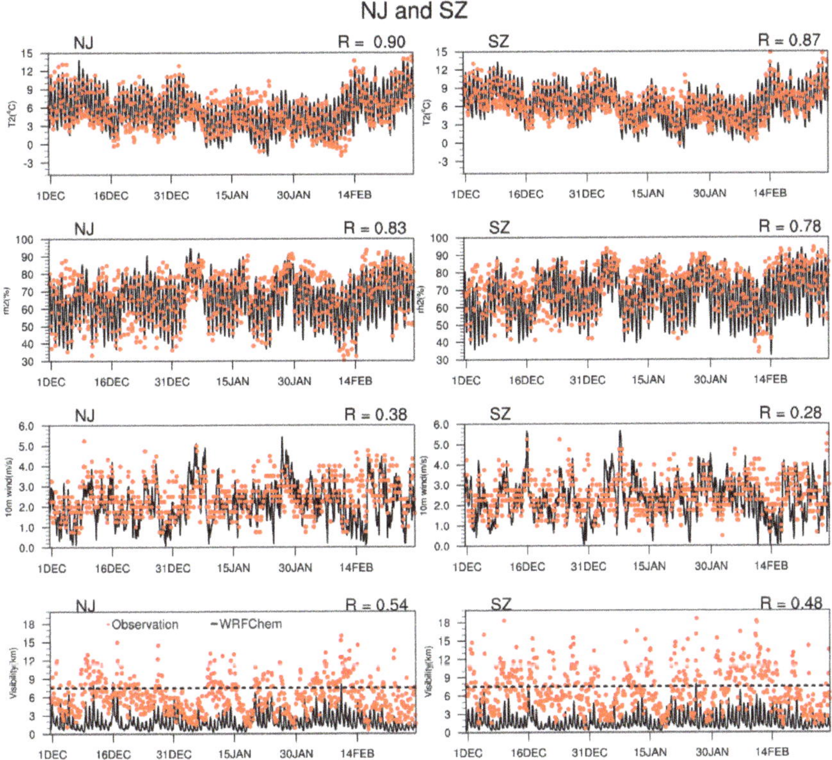

Figure 7. The temporal variation of multi-winter (2013–2017) mean 3-hourly T_2, RH_2, WS_{10}, and VIS from top to bottom, obtained from observation (red dots) and WRF-Chem simulation (black curves) in the city of Nanjing City (NJ) (left column) and Suzhou City (SZ) (right column).

Table 2. R, RMSE, MFE and MFB of the multi-winter (2013–2017) mean 3-hourly T_2, RH_2, WS_{10}, and VIS simulated by WRF-Chem model compared to the observation in the four cities of XZ, YC, NJ, and SZ.

Weather Element	City	R	RMSE	MFE	MFB
$T_2(°C)$	XZ	0.90	1.51	−0.22	0.45
	YC	0.91	1.38	0.24	0.40
	NJ	0.90	1.36	−0.16	0.14
	SZ	0.87	1.30	0.014	0.10
$RH_2(\%)$	XZ	0.78	10.50	−0.06	0.10
	YC	0.84	11.26	−0.08	0.09
	NJ	0.83	7.88	−0.01	0.07
	SZ	0.78	9.25	−0.04	0.07
$WS_{10}(m/s)$	XZ	0.36	1.01	−0.02	0.34
	YC	0.44	1.07	−0.08	0.27
	NJ	0.38	1.07	−0.08	0.25
	SZ	0.28	1.14	0.00	0.26
Visibility (VIS, km)	XZ	0.59	3.16	−0.03	0.26
	YC	0.53	6.52	0.16	0.33
	NJ	0.54	3.26	−0.24	0.34
	SZ	0.48	4.49	−0.30	0.42

3.2. WRF-Chem Simulation of Air Pollutants in Jiangsu Province

The monthly mean concentrations of SO_2, NO_2, $PM_{2.5}$, PM_{10}, CO, and O_3 in Jiangsu Province during the winters of 2013–2017 obtained from the observation and WRF-Chem simulation are shown in Figure 8. WRF-Chem has a good ability to simulate the monthly variation of pollutants in Jiangsu Province in winter. Nevertheless, exceptions of underestimating $PM_{2.5}$ and overestimating PM_{10} are also observed.

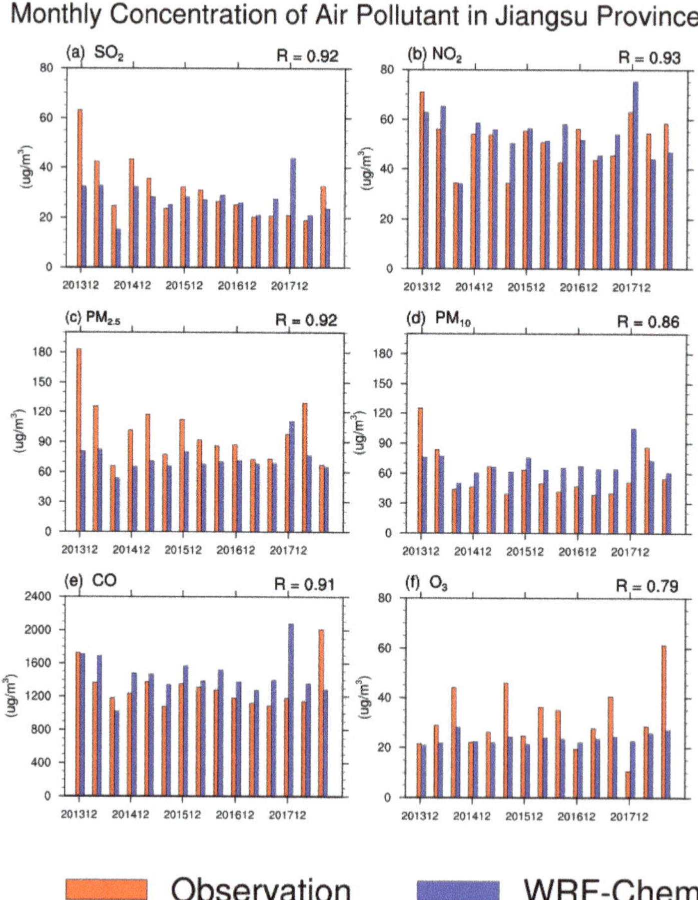

Figure 8. Comparison of monthly mean atmospheric pollutant concentrations between observation and WRF-Chem simulation in winters of 2013–2017.

Figure 9 shows the diurnal variation curves of SO_2, NO_2, $PM_{2.5}$, PM_{10}, CO, and O_3 obtained from observation and WRF-Chem simulation. SO_2, NO_2, $PM_{2.5}$, PM_{10}, and CO have similar diurnal variations. The concentration begins to increase at approximately 08:00 LST and then slowly declines in the afternoon, reaching the lowest level at approximately 15:00 LST. Then, the value increases again and becomes steady after approximately 17:00 LST. The variation is mainly due to human activities. In the rush hours of approximately 08:00 LST and 17:00 LST, automobile emissions are significantly higher, thus resulting in an increase in pollutants. The variation curve of O_3 also begins to increase at approximately 08:00 LST followed by a peak observed at 15:00 LST. Then, levels begin to decline and are relatively stable after 19:00 LST. This is because solar radiation is the main factor affecting the concentration of O_3. The test results indicate that the R values of $PM_{2.5}$, PM_{10}, CO, and O_3 obtained by WRF-Chem simulation pass the 90% confidence test (R threshold is 0.33). The simulation reasonably reproduces the daily variation characteristics of each pollutant concentration.

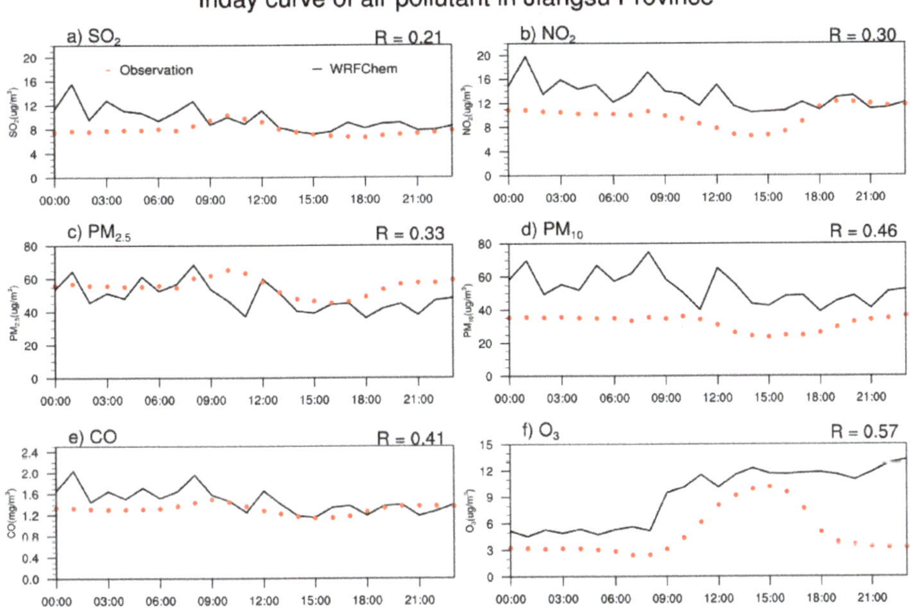

Figure 9. Diurnal variation curves of SO_2, NO_2, $PM_{2.5}$, PM_{10}, CO, and O_3 obtained from observation and WRF-Chem modulation. The R thresholds of passing 90% and 95% confidence tests are 0.33 and 0.388, respectively.

3.3. Test for Modified Visibility Based on the Neural Network Scheme

WRF-Chem shows a low ability to represent the daily visibility over Jiangsu Province. The R of the daily mean visibility between observation and inversion in winters of 2014–2017 is only 0.17 (Figure 10). Therefore, we used the neural network to modify the daily visibility series obtained from WRF-Chem inversion.

Based on previous work on the physical linkage between weather conditions and winter haze, the contribution of near-surface temperature, wind, and other meteorological factors to the variance of haze episodes can exceed 50% [1,30,31]. Moreover, Section 3.1 also proves that WRF-Chem has a strong ability to simulate these three meteorological elements. Therefore, three variables, including 2-m temperature, 10-m meridional and zonal wind speed, obtained from WRF-Chem simulation are employed for the neural network scheme. The training period is 90 days in the winter of 2013. The daily visibility series in winters of 2014–2017 are modified, as shown in Figure 10b.

Figure 10a shows the daily visibility series in the winters of 2014–2017 obtained from observation and WRF-Chem based on extinction coefficient inversion. Notably, the visibility inverted by this method is generally underestimated. This problem also occurs in the hourly series in Figure 4. After the modification of the neural network model by introducing meteorological factors, the underestimation of the inverted visibility is mostly solved. The statistical results of R, RMSE, MFE and MFB are shown in Table 3. Remarkably, the neural network significantly improves the visibility directly inverted by WRF-Chem based on the extinction coefficient. Here, R increased from 0.17 to 0.42, and RMSE/MFE/MFB decreased from 2.62/−0.26/0.33 to 1.76/−0.05/0.12.

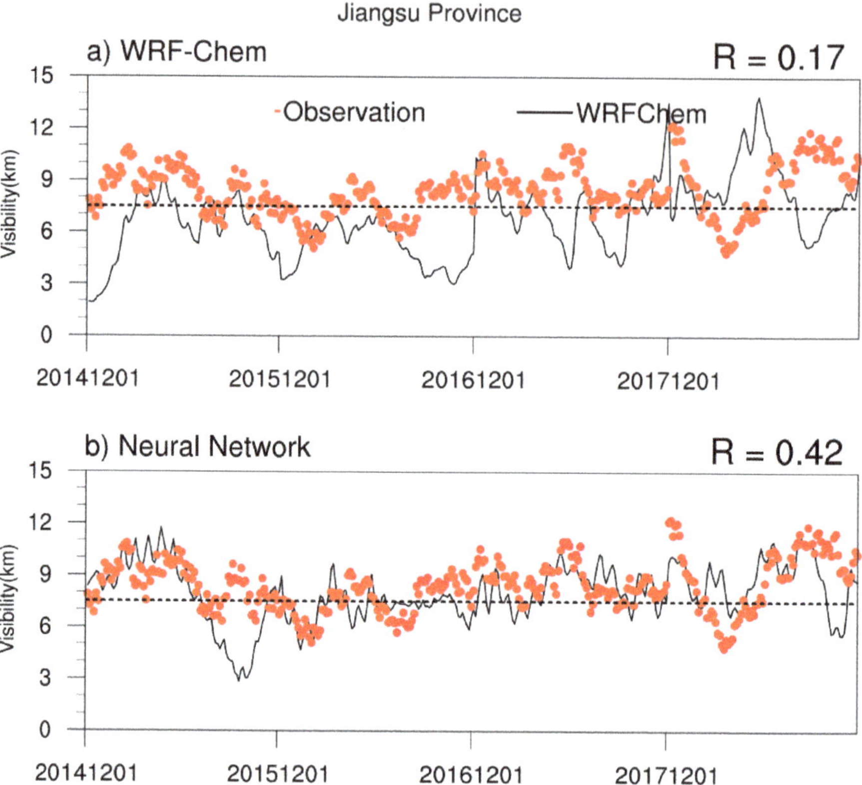

Figure 10. Time series of daily mean visibility obtained from observation and WRF-Chem model simulation corrected by the neural network in Jiangsu Province in winters of 2014–2017.

Table 3. R, RMSE, MFE and MFB of daily mean visibility (360 days by length) between observation and WRF-Chem model simulation corrected by neural network in Jiangsu Province in winters of 2014–2017.

Scheme	R	RMSE	MFE	MFB
IMPROVE	0.17	2.62	−0.26	0.33
Neural Network	0.42	1.76	−0.05	0.12

To further evaluate the correction ability of the neural network for visibility simulation, the visibility obtained from inversion is evaluated at different humidity and wind speed levels. In view of the daily series of surface relative humidity (RH) in winters of 2014–2017, there are only two days with RH less than 60% and zero days more than 90%. Therefore, the observed RH is divided into four grades, namely, RH < 70%, 70% ≤ RH < 75%, 75% ≤ RH < 80% and RH ≥ 80%. Then, the MFE and MFB of atmospheric visibility simulated by the two schemes in each grade are calculated. Similarly, wind speed (WS) is divided into three grades, namely WS < 2 m/s, 2 m/s ≤ WS < 2.5 m/s, and WS > 2.5 m/s, and the corresponding MFE and MFB are also computed. As noted in Figure 11, the visibility inversion performance of the IMPROVE extinction coefficient formula is unsatisfactory under low RH, and the visibility is significantly underestimated. However, its performance improves as RH increases. The performance of the visibility inversion scheme by the neural network is better than the IMPROVE method at each humidity level. In particular, the IMPROVE method is significantly improved under high wind speeds. For different level wind speeds, the greater the WS is, the worse the performance of the IMPROVE method. Moreover, the visibility is significantly underestimated under high WS. The performance of visibility inversion by the neural network under high WS is much better than the IMPROVE method.

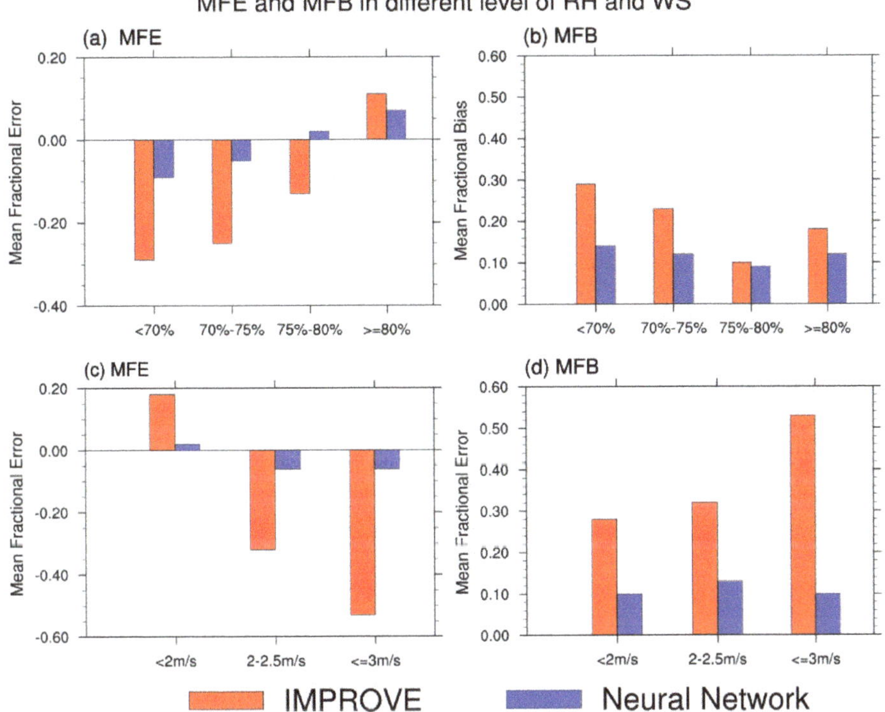

Figure 11. MFE and MFB derived from visibility of different surface relative humidity and wind speed levels by Interagency Monitoring of Protected Visual Environments (IMPROVE) and neural network observations.

4. Conclusions and Discussions

In this paper, high-resolution downscaling simulation of meteorological elements and air quality in Jiangsu Province in winters of 2013–2017 was performed using the air quality model WRF-Chem. The simulation was conducted at a spatial resolution of 18 km and a temporal resolution of 1 h. The inverted daily visibility of Jiangsu Province in winter was also calculated. The results show that WRF-Chem can accurately simulate the meteorological elements and air quality in Jiangsu Province and the four representative cities of XZ, YC, NJ, and SZ. Regarding air pollutants, NO_2, PM_{10}, and O_3 are overestimated, while $PM_{2.5}$ is underestimated. To some extent, the overestimation of the air pollutants leads to the low daily visibility in winter generated from the visibility inversion scheme based on the IMPROVE extinction coefficient calculation formula.

The IMPROVE extinction coefficient calculation formula mainly takes the factors of atmospheric chemical composition into account, while no meteorological factors are considered except RH. However, studies show that near-surface temperature, wind field, and other meteorological factors can contribute to greater than 50% to the variance of haze episodes. Therefore, meteorological factors that have obvious influence on visibility were employed, including 2-m air temperature, 10-m meridional and zonal wind, to improve the visibility inversion scheme based on the extinction coefficient. Considering the complex and nonlinear relationship between meteorological factors and visibility, a deep learning method, namely a neural network, was adopted. By self-modifying and modeling through machine learning, the neural network can create a modified model suitable for the simulation. The results indicate that the neural network can significantly improve the visibility inversion based on the extinction coefficient formula to solve the issue of visibility underestimation. In addition, under different RH and WS levels, the neural network has significant improvements for the IMPROVE scheme. The performance of the IMPROVE scheme to invert visibility is relatively poor under low RH or high WS, which are conducive to pollutant diffusion. Thus, the visibility is significantly underestimated. The problem is mainly caused by only considering the extinction impact of aerosol factors but leaving out meteorological factors. In short, the inverted visibility under low RH and high WS is significantly improved after introducing meteorological factors by multilayer neural network modeling.

This study fully examined the ability of the WRF-Chem model in simulating winter visibility and related meteorological variables in Jiangsu Province in eastern China. WRF-Chem reliably reproduces winter haze-related variables, especially after applying the multilayer neural network scheme. Based on this work, a dynamic downscaling study to predict the winter visibility in Jiangsu Province using WRF-Chem and the neural network scheme is on-going. Considering the possible effect of the emission and model resolution, further refinement in the emission and model resolution could help improve the simulation and prediction of regional visibility.

Author Contributions: Conceptualization, H.W., Y.Z. and P.Z.; methodology, H.W., Y.Z. and P.Z.; software, P.Z.; validation, Y.Z., P.Z. and D.L.; formal analysis, P.Z. and Y.Z.; investigation, P.Z.; resources, P.Z. and D.L.; data curation, P.Z. and D.L.; writing—original draft preparation, P.Z.; writing—review and editing, H.W. and Y.Z.; visualization, P.Z.; supervision, H.W.; project administration, Y.Z. and P.Z.; funding acquisition, Y.Z. and P.Z. All authors have read and agreed to the published version of the manuscript.

Funding: This research was funded by the National Key R&D Program of China (2017YFC1502304), and National Natural Science Foundation of China (41805001, and 41675083).

Conflicts of Interest: The authors declare no conflict of interest.

References

1. Guo, Y.; Li, S.; Zhou, Y.; Chen, L.; Chen, S.; Yao, T.; Wu, S. The effect of air pollution on human physiological function in China: A longitudinal study. *Lancet* **2015**, *S31*, 386. [CrossRef]
2. Zanobetti, A.; Schwartz, J. The Effect of Fine and Coarse Particulate Air Pollution on Mortality: A National Analysis. *Environ. Health Perspect.* **2009**, *117*, 898–903. [CrossRef] [PubMed]
3. Zhang, A.; Zhong, L.; Xu, Y.; Wang, H.; Dang, L. Tourists' Perception of Haze Pollution and the Potential Impacts on Travel: Reshaping the Features of Tourism Seasonality in Beijing, China. *Sustainability* **2015**, *7*, 2397–2414. [CrossRef]
4. Stevens, C.J.; Dise, N.B.; Mountford, J.O.; Gowing, D.J. Impact of nitrogen deposition on the species richness of grasslands. *Science* **2004**, *303*, 1876–1879. [CrossRef] [PubMed]
5. Zhang, R.H.; Li, Q.; Zhang, R.N. Meteorological conditions for the persistent severe fog and haze event over eastern China in January 2013. *Sci. China Earth Sci.* **2014**, *57*, 26–35.
6. Chen, H.P.; Wang, H.J. Haze days in North China and the associated atmospheric circulations based on daily visibility data from 1960 to 2012. *J. Geophys. Res. Atmos.* **2015**, *120*, 5895–5909. [CrossRef]
7. Wang, H.J.; Chen, H.P.; Liu, J.P. Arctic sea ice decline intensified haze pollution in eastern China. *Atmos. Oceanic Sci. Lett.* **2015**, *8*, 1–9.
8. Li, S.L.; Han, Z.; Chen, H.P. A comparison of the effects of interannual Arctic sea ice loss and ENSO on winter haze days: Observational analyses and AGCM simulations. *J. Meteorol. Res.* **2017**, *31*, 820–833. [CrossRef]
9. Liu, D.; Wei, J.; Kang, Z.; Yan, W.; Cao, L.; Chen, H. Urbanization and Industrialization Effects on Haze in China: Take Jinagsu for Example. In Proceedings of the 19th EGU General Assembly, EGU2017, Vienna, Austria, 23–28 April 2017; 2017; p. 18493.
10. Gao, X.; Xu, Y.; Zhao, Z. Jeremy Impacts of horizontal resolution and topography on the numerical simulation of East Asian precipitation (in Chinese). *Chin. J. Atmos. Sci.* **2006**, *30*, 185–192.
11. Yu, E.T.; Sun, J.Q.; Chen, H.P.; Xiang, W. Evaluation of a high-resolution historical simulation over China: Climatology and extremes. *Clim. Dyn.* **2015**, *45*, 2013–2031. [CrossRef]

12. Wang, J.; Wang, H.J.; Hong, Y. Comparison of satellite-estimated and model-forecasted rainfall data during a deadly debris-flow event in Zhouqu, Northwest China. *Atmos. Ocean. Sci. Lett.* **2016**, *9*, 139–145. [CrossRef]

13. Wei, W.; Lv, Z.F.; Li, Y.; Wang, L.; Cheng, S. A WRF-Chem model study of the impact of VOCs emission of a huge petro-chemical industrial zone on the summertime ozone in Beijing, China. *Atmos. Environ.* **2017**, *175*, 44–53. [CrossRef]

14. Zhou, G.; Xu, J.; Xie, Y.; Chang, L.; Gao, W.; Gu, Y. Numerical air quality forecasting over eastern China: An operational application of WRF-Chem. *Atmos. Environ.* **2017**, *153*, 94–108. [CrossRef]

15. Song, H.; Wang, K.; Zhang, Y.; Hong, C.; Zhou, S. Simulation and evaluation of dust emissions with WRF-Chem (v3.7.1) and its relationship to the changing climate over East Asia from 1980 to 2015. *Atmos. Environ.* **2017**, *167*, 511–522. [CrossRef]

16. Peng, H.; Liu, D.; Zhou, B.; Su, Y.; Wu, J.; Shen, H. Boundary-Layer Characteristics of Persistent Regional Haze Events and Heavy Haze Days in Eastern China. *Adv. Meteorol.* **2016**, *2016*, 1–23.

17. Dai, Z.; Liu, D.; Yu, K.; Cao, L.; Jiang, Y. Meteorological Variables and Synoptic Patterns Associated with Air Pollutions in Eastern China during 2013–2018. *Int. J. Environ. Res. Public Health* **2020**, *17*, 2528. [CrossRef]

18. Li, Z.H.; Liu, D.Y.; Yan, W.L.; Wang, H.B.; Zhu, C.Y.; Zhu, Y.Y.; Zu, F. Dense fog burst reinforcement over Eastern China: A review. *Atmos. Res.* **2019**, *230*, 104639. [CrossRef]

19. Liu, Y.Z.; Wang, B.; Zhu, Q.Z.; Luo, R.; Wu, C.Q.; Jia, R. Dominant synoptic patterns and their relationships with PM2.5 pollution in winter over the Beijing-Tianjin-Hebei and Yangtze River Delta Regions. *J. Meteor. Res.* **2019**, *33*, 765–776. [CrossRef]

20. Relvas, H.; Miranda, A.I. An urban air quality modeling system to support decision-making: Design and implementation. *Air. Qual. Atmos. Health* **2018**, *11*, 815–824. [CrossRef]

21. Claudio, C.; Giovanna, F.; Enrico, P.; Marialuisa, V. Neuro-fuzzy and neural network systems for air quality control. *Atmos. Environ.* **2009**, *43*, 4811–4821.

22. Grell, G.A.; Peckham, S.E.; Schmitz, R. Fully coupled "online" chemistry within the WRF model. *Atmospheric Environment.* **2005**, *2005*, 6957–6975. [CrossRef]

23. Kain, J.S. The Kain-Fritsch convective parameterization: An update. *J. Appl. Meteorol. Climatol.* **2004**, *43*, 170–181. [CrossRef]

24. Chen, S.-H.; Sun, W.-Y. A One-dimensional Time Dependent Cloud Model. *J. Meteorol. Soc. Japan. Ser. II.* **2002**, *80*, 99–118. [CrossRef]

25. Baek, S. A revised radiation package of G-packed McICA and two-stream approximation: Performance evaluation in a global weather forecasting model. *J. Adv. Modeling Earth Syst.* **2017**, *9*, 3. [CrossRef]

26. Hong, S.Y.; Noh, Y.; Dudhia, J. A new vertical diffusion package with an explicit treatment of entrainment processes. *Mon. Weather Rev.* **2006**, *134*, 2318–2341. [CrossRef]

27. Chen, F.; Dudhia, J. Coupling an advanced land surface hydrology model with the Penn State–NCARMM5 modeling system. Part I: Model implementation and sensitivity. *Mon. Weather Rev.* **2001**, *129*, 569–585. [CrossRef]

28. Jose, R.S.; Perez, J.L.; Balzarini, A.; Baro, R.; Curci, G.; Forkel, R. Sensitivity of feedback effects in CBMZ/MOSAIC chemical mechanism. *Atmos. Environ.* **2015**, *115*, 646–656. [CrossRef]

29. Fast, J.D.; Gustafson, J.; Zaveri, R.C. Evolution of ozone, particulates, and aerosol direct forcing in an urban area using a new fully-coupled meteorology, chemistry, and aerosol model. *J. Geophys. Res.* **2006**, *2006*, D21305. [CrossRef]

30. Chapman, E.C.; Gustafson, W.I.; Easter, R.C. Coupling aerosol-cloud-radiative processes in the WRF-Chem model: Investgating the radiative impact of elevated point source. *Atmos. Chem. Phys.* **2009**, *9*, 945–964. [CrossRef]

31. Zhang, Y.; Wen, X.Y.; Jang, C.J. Simulating chemistry–aerosol–cloud–radiation–climate feedbacks over the continental U.S. using the online-coupled Weather Research Forecasting Model with chemistry (WRF/Chem). *Atmos. Environ.* **2010**, *44*, 3568–3582. [CrossRef]

32. Zhang, Q.; Streets, D.G.; Carmichael, G.R.; He, K.B.; Huo, H.; Kannari, A.; Klimont, Z.; Park, I.S.; Reddy, S.; Fu, J.S.; et al. Asian emissions in 2006 for the NASA INTEXB mission. *Atmos. Chem. Phys.* **2009**, *9*, 5131–5153. [CrossRef]

33. Li, K.; Liao, H.; Mao, Y.H.; David, A.R. Source sector and region contributions to concentration and direct radiative forcing of black carbon in China. *Atmos. Environ.* **2015**, *124*, 351–366. [CrossRef]

34. Li, M.; Zhang, Q.; Streets, D.G.; He, K.B.; Cheng, Y.F.; Emmons, L.K.; Huo, H.; Kang, S.C.; Lu, Z.; Shao, M.; et al. Mapping Asian anthropogenic emissions of non-methane volatile organic compounds to multiple chemical mechanisms. *Atmos. Chem. Phys.* **2014**, *14*, 5617–5638. [CrossRef]
35. Zheng, B.; Huo, H.; Zhang, Q.; Yao, Z.L.; Wang, X.T.; Yang, X.F.; Liu, X.; He, K.B. High-resolution mapping of vehicle emissions in China in 2008. *Atmos. Chem. Phys.* **2014**, *14*, 9787–9805. [CrossRef]
36. Sisler, J.F.; Malm, W.C. Interpretation of trends of $PM_{2.5}$ and reconstructed visibility from the IMPROVE network. *J. Air Waste Manag. Assoc.* **2000**, *50*, 775–789. [CrossRef] [PubMed]
37. Chow, J.C.; Bachmann, J.D.; Wierman, S.S.G.; Chow, J.C.; Bachmann, J.D.; Wierman, S.S.G.; Mathai, C.V.; Malm, W.C.; &White, W.H. Vsibility: Science and regulation. *J. Air Waste Manag. Assoc.* **2002**, *52*, 973–999. [CrossRef] [PubMed]
38. Pitchford, M.; Main, W.; Schichtel, B. Revised algorithm for estimating light extinction from IMPROVE particle speciation data. *J. Air Waste Manag. Assoc.* **2007**, *57*, 1326–1336. [CrossRef]
39. Qiu, Y.; Ma, Z.; Li, K. A modeling study of the peroxyacetyl nitrate (PAN) during a wintertime haze event in Beijing, China. *Sci. Total Environ.* **2019**, *650*, 1944–1953. [CrossRef]
40. Zhang, Z.; Gong, D.; Mao, R.; Kim S., J.; Xu, J.; Zhao, X. Cause and predictability for the severe haze pollution in downtown Beijing in November–December 2015. *Sci. Total Environ.* **2017**, *592*, 627–638. [CrossRef]
41. Chollet, F. *Deep Learning with Python*; Manning Publications: Westampton, NJ, USA, 2018.
42. Boylan, J.W.; Russell, A.G. PM and light extinction model performance metrics, goals, and criteria for three-dimensional air quality models. *Atmos. Environ.* **2006**, *40*, 4946–4959. [CrossRef]

Article

Model Inter-Comparison for $PM_{2.5}$ Components over urban Areas in Japan in the J-STREAM Framework

Kazuyo Yamaji [1,*], Satoru Chatani [2], Syuichi Itahashi [3], Masahiko Saito [4], Masayuki Takigawa [5], Tazuko Morikawa [6], Isao Kanda [7], Yukako Miya [8], Hiroaki Komatsu [9], Tatsuya Sakurai [10], Yu Morino [2], Kyo Kitayama [2], Tatsuya Nagashima [2], Hikari Shimadera [11], Katsushige Uranishi [11], Yuzuru Fujiwara [12], Tomoaki Hashimoto [12], Kengo Sudo [13], Takeshi Misaki [1,14] and Hiroshi Hayami [3]

[1] Graduate School of Maritime Sciences, Kobe University, Kobe, Hyogo 658-0022, Japan; misaki@tepsco.co.jp
[2] National Institute for Environmental Studies, Tsukuba, Ibaraki 305-8506, Japan; chatani.satoru@nies.go.jp (S.C.); morino.yu@nies.go.jp (Y.M.); kitayama.kyo@nies.go.jp (K.K.); nagashima.tatsuya@nies.go.jp (T.N.)
[3] Central Research Institute of Electric Power Industry, Abiko, Chiba 270-1194, Japan; isyuichi@criepi.denken.or.jp (S.I.); haya@criepi.denken.or.jp (H.H.)
[4] Ehime University, Matsuyama, Ehime 790-8566, Japan; msaito@agr.ehime-u.ac.jp
[5] Japan Agency for Marine—Earth Science and Technology, Yokohama, Kanagawa 236-0001, Japan; takigawa@jamstec.go.jp
[6] Japan Automobile Research Institute, Tsukuba, Ibaraki 305-0822, Japan; tmorikaw@jari.or.jp
[7] Japan Meteorological Corporation, Osaka 556-0021, Japan; isao.kanda@n-kishou.co.jp
[8] Japan Weather Association, Toshima, Tokyo 170-6055, Japan; yukako@jwa.or.jp
[9] Kanagawa Environmental Research Center, Hiratsuka, Kanagawa 254-0014, Japan; komatsu.amti@pref.kanagawa.jp
[10] School of Science and Engineering, Meisei University, Hino, Tokyo 191-8506, Japan; tatsuya.sakurai@meisei-u.ac.jp
[11] Graduate School of Engineering, Osaka University, Suita, Osaka 565-0871, Japan; shimadera@see.eng.osaka-u.ac.jp (H.S.); k.uranishi2@gmail.com (K.U.)
[12] Suuri-Keikaku, Chiyoda, Tokyo 101-0003, Japan; fujiwara@sur.co.jp (Y.F.); hashimoto_tomoaki@sur.co.jp (T.H.)
[13] Graduate School of Environmental Studies, Nagoya University, Nagoya, Aichi 464-8601, Japan; kengo@nagoya-u.jp
[14] Tokyo Electric Power Services Co., Ltd., Koto, Tokyo 135-0062, Japan
[*] Correspondence: kazuyo@maritime.kobe-u.ac.jp; Tel.: +81-78-431-6282

Received: 14 January 2020; Accepted: 19 February 2020; Published: 25 February 2020

Abstract: A model inter-comparison of secondary pollutant simulations over urban areas in Japan, the first phase of Japan's study for reference air quality modeling (J-STREAM Phase I), was conducted using 32 model settings. Simulated hourly concentrations of nitric oxide (NO) and nitrogen dioxide (NO_2), which are primary pollutant precursors of particulate matter with a diameter of 2.5 μm or less ($PM_{2.5}$), showed good agreement with the observed concentrations, but most of the simulated hourly sulfur oxide (SO_2) concentrations were much higher than the observations. Simulated concentrations of $PM_{2.5}$ and its components were compared to daily observed concentrations by using the filter pack method at selected ambient air pollution monitoring stations (AAPMSs) for each season. In general, most models showed good agreement with the observed total $PM_{2.5}$ mass concentration levels in each season and provided goal or criteria levels of model ensemble statistics in warmer seasons. The good performances of these models were associated with the simulated reproducibility of some dominant components, sulfates (SO_4^{2-}) and ammonium (NH_4^+). The other simulated $PM_{2.5}$ components, i.e., nitrates (NO_3^-), elemental carbon (EC), and organic carbon (OC), often show clear deviations from the observations. The considerable underestimations (approximately 30 μg/m^3 for total $PM_{2.5}$) of all participant models found on heavily polluted days with approximately 40–50 μg/m^3 for total $PM_{2.5}$ indicated some problems in the simulated local meteorology such as the atmospheric stability.

This model inter-comparison suggests that these deviations may be owing to a need for further improvements both in the emission inventories and additional formation pathways in chemical transport models, and meteorological conditions also require improvement to simulate elevated atmospheric pollutants. Additional accumulated observations are likely needed to further evaluate the simulated concentrations and improve the model performance.

Keywords: PM$_{2.5}$; PM$_{2.5}$ components; three-dimensional chemical transport model; model inter-comparison; urban scale; secondary particles

1. Introduction

Particulate matter (PM) consists of a complex mixture of solid and liquid particles of organic and inorganic substances suspended in the atmosphere. The major components of PM are sulfates (SO_4^{2-}), nitrates (NO_3^-), ammonium (NH_4^+), sodium chloride (NaCl), black carbon (BC) or elemental carbon (EC), organic carbon (OC), mineral dust, and water. The PMs with the greatest negative health effects are those with a diameter of 2.5 μm or less (PM$_{2.5}$), which can penetrate and lodge deep inside the lungs [1]. Some PMs are also climate-dependent, known as short-lived climate pollutants (SCLPs) [2]. Warming due to sunlight absorption (e.g., BC) and cooling due to sunlight scattering (e.g., SO_4^{2-}) directly affect radiative forcing in the earth's climate system. Additionally, water-soluble PMs affect the regional climate system interacting with cloud microphysics. These radiative and microphysical interactions can induce changes in regional precipitation and atmospheric circulation patterns. Some PM$_{2.5}$ particles are directly emitted from natural sources and human activities, while others are formed through complex oxidation reactions and particle agglomeration. Combining the regional three-dimensional chemical transport model (CTM) with comprehensive particulate formations may be a useful tool for understanding the detailed behavior of short-lived PM$_{2.5}$ components in the atmosphere.

Recently, PM$_{2.5}$ air quality has been improved in East Asian countries, e.g., China [3] and Japan [4]; however, PM$_{2.5}$ concentrations at Japanese air pollution monitoring stations (APMSs) have not met yet the environmental quality standard, defined as 15 μg/m^3 for the annual PM$_{2.5}$ mean and 35 μg/m^3 for 24-h PM$_{2.5}$ mean, or the World Health Organization (WHO) air quality guidelines, with corresponding values of 10 and 35 μg/m^3. An established reference regional CTM system should be applied to design effective PM$_{2.5}$ control strategies [5]. However, accurately reproducing or predicting the concentrations and distributions of PM$_{2.5}$ and its relevant substances remains challenging, due to inaccurate emission inventories, poorly represented initial and boundary conditions, imperfect physical, dynamical, and chemical parameterizations, and limited observations for verification, as noted for previous Asian scale model inter-comparisons, i.e., the model inter-comparison study for Asia (MICS-Asia) series [6–9] and the urban air quality model inter-comparison study in Japan (UMICS) series [10–12].

A model inter-comparison framework, Japan's study for reference air quality modeling (J-STREAM), was designed to establish a reference regional CTM system to consider strategies for reducing PM$_{2.5}$ and its relevant substances [5]. In this paper, the capacities of participant models for J-STREAM to simulate PM$_{2.5}$ and its components were evaluated for two urban areas in Japan in each season. The model improvements will be discussed based on the inter-model differences.

2. Methodology

2.1. Framework of J-STREAM Phase I

A model inter-comparison project in Japan, J-STREAM, was initiated in 2016. One aim of J-STREAM is to investigate differences in simulated concentrations of secondary atmospheric pollutants such as PM$_{2.5}$ components and ozone over urban areas in Japan due to differences between model frames and/or model settings, including boundary and inputted conditions and physical and chemical mechanisms.

Detailed model settings are described below. Furthermore, these including an introduction of J-STREAM can be found in previous research for the overview [5] and the performance on ozone [13].

The main target of the first phase of J-STREAM (J-STREAM Phase I) is to evaluate the general performances of participant models on secondary atmospheric concentrations over urban areas in Japan. Daily concentrations of $PM_{2.5}$ components in each season among others were treated as subjects of evaluation in this paper. The enhanced simulation periods of J-STREAM Phase I were the spring of 2013, 27 April–26 May 2013, the summer of 2013, 12 July–10 August 2013, the autumn of 2013, 11 October–9 November 2013, and the winter of 2014, 10 January–8 February 2014, which corresponded to the seasonal periods of the national observation frame for $PM_{2.5}$ components (Table 1). The detailed evaluations and additional experiments for individual participant models can be found in [14].

Table 1. Dates of enhanced simulation periods for model evaluations including a simulation spin-up. Updated from the overview of Japan's study for reference air quality modeling (J-STREAM) [5].

Season	Dates
Spring 2013	27 April–26 May 2013
Summer 2013	12 July–10 August 2013
Autumn 2013	11 October–9 November 2013
Winter 2014	10 January–8 February 2014

Four nested model domains, d01, d02, d03, and d04, on a Lambert conformal map projection were employed in the J-STREAM project [5]. The finest domains, d03 and d04, with a 5 × 5 km grid, cover the major city clusters in western Japan, including Osaka, Kobe, Kyoto, and Nagoya, and the Tokyo metropolitan area, respectively. Simulated concentrations in the d03 and d04 domains were used for model evaluations, and the results are discussed in following sections. Figure 1 shows the d03 and d04 domains, including the locations of ambient APMSs (AAPMSs), for which simulated concentrations were evaluated via comparisons with observations.

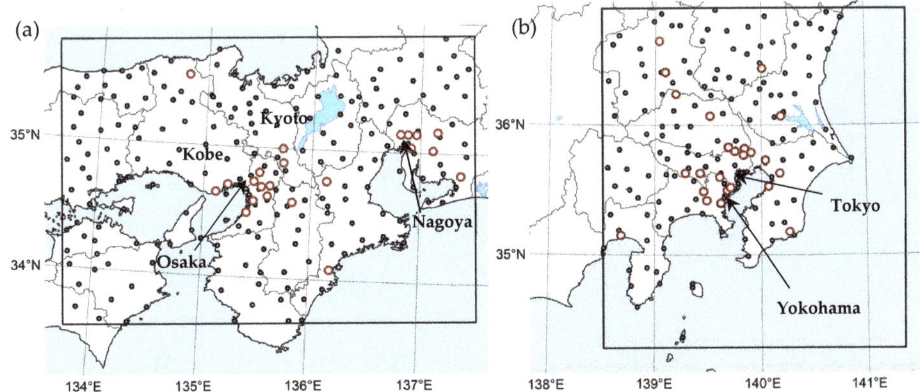

Figure 1. Finest model domains for J-STREAM, d03 (**a**) and d04 (**b**). d03 covers major city clusters located in western Japan including Osaka, Kobe, Kyoto, and Nagoya. d04 covers the Tokyo metropolitan area. The red circles indicate the locations of ambient air pollution monitoring stations (AAPMSs). The black circles indicate the locations of the meteorological observation stations of the Japan Meteorological Agency (JMA).

2.2. Baseline Meteorological Model Configurations

The baseline meteorological simulation for J-STREAM Phase I was performed by the Weather Research and Forecasting (WRF) model, using the Advanced Research WRF (ARW) Version 3.7.1 [15].

The WRF inputted data were acquired from the National Centers for Environmental Prediction Final Operational Model Global Tropospheric Analyses (ds083.2) with a 1 × 1 degree resolution [16] and the Real-Time, Global Sea Surface Temperature High-Resolution (RTG_SST_HR) analysis with a 1/12 × 1/12 degree resolution [17] and a temporal resolution of 6 h for the initial and boundary conditions. The horizontal configurations of the one-way nested model domains, d01, d02, d03, and d04, are 220 × 170 grids with a 45-km horizontal resolution, 154 × 160 grids with a 15-km resolution, 82 × 61 grids with a 5-km resolution, and 64 × 70 grids with a 5-km resolution, respectively. The vertical grid structure consists of 31 layers from the surface to the model top (100 hPa). Five grids were trimmed off each of the four lateral boundaries for the offline CTMs. The physics parameterizations applied in this model included the WRF Single-Moment 5-class scheme [18], the Radiative Transfer Model (RRTM) [19] for a longwave radiation scheme, the Dudhia scheme [20] for a shortwave radiation scheme, the Noah Land Surface Model [21], the Mellor-Yamada Nakanishi and Niino surface layer scheme level 2.5 [22], and the Kain-Fritsch convective parameterization [23] for d01 and d02. No convection parameterization was used for the 5-km domains. The grid-nudging four-dimensional data assimilation technique was employed for wind, temperature, and water vapor from level 11 (approximately 2 km) to the top of the model at 100 hPa with the nudging coefficients of 1.0×10^{-4} and 0.5×10^{-5} s^{-1} for d01 and d02, respectively. Most of the participant CTMs employed baseline meteorological fields, while others employed the meteorology based on different model settings. The differences in the model settings in some participant models are described in Section 2.3.

The baseline meteorological fields were compared with hourly observations of the Japan Meteorological Agency (JMA) for the observation stations within d03 and d04 (Figure 1) over four seasons: the spring of 2013 (11–26 May 2013), the summer of 2013 (27 July–10 August 2013), the autumn of 2013 (25 October–9 November 2013), and the winter of 2014 (24 January–7 February 2014) (Figures 2 and 3). The hourly observed and modeled meteorological variables were averaged for all meteorological observatories for each domain.

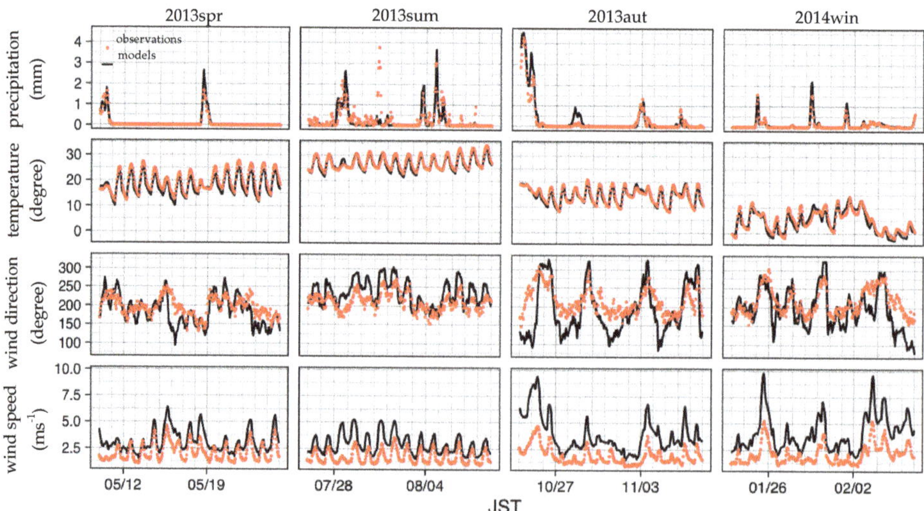

Figure 2. Spatially averaged precipitation, temperature, wind direction, and wind speed over four seasons. These results are based on hourly observed values and simulated at all JMA stations for d03.

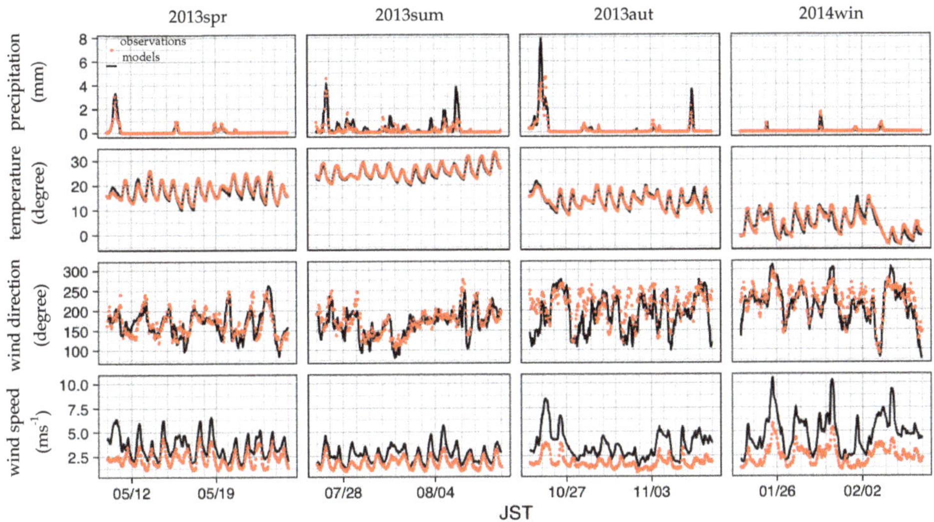

Figure 3. Spatially averaged precipitation, temperature, wind direction, and wind speed over the four seasons. These results are based on hourly observed values and simulated at all JMA stations for d04.

The WRF using the baseline setting can generally simulate the observed meteorological conditions well. Meanwhile, WRF tended to overestimate the observed wind speeds. This was likely affected by the sparse horizontal resolution and coarse land information. The simulation performance of wind patterns was slightly better for d04 than that for d03 (Figures 2 and 3). However, simulated precipitation timing and their amounts were consistent with the observations (Figures 2 and 3).

2.3. Chemical Transport Model Configurations

A total of 32 simulations were performed using three types of regional CTMs in J-STREAM Phase I: Community Multiscale Air Quality (CMAQ) [24], Comprehensive Air quality Model with eXtensions (CAMx) [25], and Weather Research and Forecasting-Chemistry (WRF-Chem) [26]. Table 2 presents the configurations of the employed models. All participants conducted J-STREAM simulation under their own usual simulation conditions. The CMAQ group (M01–M28) included several versions, i.e., chemical mechanisms: Statewide Air Pollution Research Center mechanism (SAPRC) 99 [27], SAPRC07 [28], Carbon Bond (CB) 05 [29], and Regional Atmospheric Chemistry Mechanism (RACM) 2 [30], and three types of CMAQ aerosol calculation techniques [31]: aero5, aero6, and aero6 with the volatility basis set (VBS) approach [32]. These CMAQ aerosol calculation techniques employed ISORROPIA Version 1 [33,34] as an aerosol thermodynamic model and the second version of ISORROPIA (ISORROPIA Version 2) for updating the crustal species thermodynamics, the speciation schemes, and the SO_4^{2-} formation pathway [35] for versions after 5.0. The basic techniques of aero5 and aero6 include secondary organic aerosol (SOA) formation processes based on empirical parameters for SOA yields [36]. Major or minor updates were reflected in the chemical and aerosol mechanisms in the later versions. One CAMx model (M29) applied in J-STREAM used the SAPRC07 chemistry and the coarse and fine aerosol scheme treating both static coarse and fine mode aerosols [37]. The WRF-Chem group (M30–M32) included two Versions (3.8.1 and 3.7.1) that employed RADM2: the aerosol module of the Modal Aerosol Dynamics Model for Europe (MADE) [38] and the SOA Model (SORGAM) [39].

As described in detail in an overview article on J-STREAM [5], participants were requested to run CTM simulations during the enhanced target periods of four seasons for d03 or d04. As shown in Table 2, the simulations for some participant models began at d01 (M02, M03, M07–M15, M20, and M30–M32), but others began from the more inner domains. Fifteen participant models (M01–07, M14,

M15, M21–M24, M26, M29, and M30) submitted their results for both domains for all four seasons, but the other participants submitted results for only selected seasons, with the highest number of model results for the summer of 2013.

Initial concentrations on the first day of each season and boundary concentrations throughout the entire target period were generated in the simulation using the M15 setting via CMAQ Version 5.0.2 with the SAPRC07–aero6 mechanisms for d01 and d02. Boundary concentrations for d01 of M15 were obtained from results for a chemical atmospheric general circulation model designed for studying atmospheric environment and radiative forcing, CHASER [40] for the Hemispheric Transport of Air Pollution (HTAP) Version 2 [41]. In J-STREAM Phase I, model-ready mosaic emission data corresponding to all participant chemical–aerosol mechanisms involving multiple emission inventories and results from an emission model for biogenic volatile organic compounds were provided: HTAP Version 2.2 [42] and Global Fire Emissions Database Version 4.1 [43] for Asian anthropogenic emissions, the Japan Auto–Oil Program (JATOP) emission inventory database (JEI-DB) [44], the updated JEI–DB [5], and Sasakawa Peace Foundation emissions for ships for Japanese anthropogenic emissions, volcanic emission data from Aerosol Comparisons between Observations and Models (AeroCom) [45] and JMA [46], and estimations obtained by using Model of Emissions of Gases and Aerosols from Nature Version 2.1 [47]. Most participant CTMs used model-ready input data; however, some participant CTMs performed simulations in their own emission frames. M03 used EAGrid2010-JAPAN [48], and M20 and M27 used EAGrid2000-JAPAN [49] for anthropogenic emissions in Japan. For the Asian scale anthropogenic emissions, M20 employed NASA INTEX-B [50] instead of HTAP Version 2.2. Additionally, some CTMs employed different emission injection heights. The Model for Ozone and Related chemical Tracers Version 4 (MOZART-4) [51], for instance, was used as boundary conditions in some model settings.

As mentioned in Section 2.2., most of the participants employed the baseline meteorological fields; however, other CTMs (M07, M20) used WRF-ARW outputs based on their own conditions, including physical options, parameterizations, and a fine input meteorological analysis data, which is the grid point value derived from the mesoscale model (GPV MSM) data by JMA.

Table 2. Configurations of participant chemical transport models (CTMs) submitted for J-STREAM Phase I, updated from an overview of J-STREAM [5].

ID	Model	Version	Chemical Mechanism	Aerosol Module	Photolysis	Simulation [1]				Emis [2]	BCON [3]	Met [4]	Submitted [5]	
						d01	d02	d03	d04				d03	d04
M01	CMAQ	5.2	CB05	aero6	inline			o	o	o	o	o	o	o
M02	CMAQ	5.1	SAPRC07	aero6	inline	o	o	o	o	o	o	o	o	o
M03	CMAQ	5.1	SAPRC07	aero6	inline	o	o	o	o	E1	o	o	o	o
M04	CMAQ	5.1	SAPRC07	aero6	inline			o	o	o	o	o	o	o
M05	CMAQ	5.1	SAPRC07	aero6	inline			o	o	o	o	o	o	o
M06	CMAQ	5.1	SAPRC07	aero6	table			o	o	o	o	o	o	o
M07	CMAQ	5.0.2	SAPRC07	aero6	inline	o	o	o	o	o	M	W	o	o
M08	CMAQ	5.0.2	SAPRC07	aero6	inline	o	o	o	o	o	o	o	o	o
M09	CMAQ	5.0.2	CB05	aero6	inline	o	o	o	o	o	o	o	o	o
M10	CMAQ	5.0.2	CB05	aero6	inline	o	o	o	o	o	o	o	su	su
M11	CMAQ	5.0.2	CB05	aero6vbs	inline	o	o	o	o	o	o	o	su	su
M12	CMAQ	5.0.2	RACM2	aero6	inline	o	o	o	o	o	o	o	o	o
M13	CMAQ	5.0.2	SAPRC99	aero5	inline	o	o	o	o	o	o	o	o	o
M14	CMAQ	5.0.2	SAPRC07	aero6	inline	o	o	o	o	o	o	o	o	o
M15	CMAQ	5.0.2	SAPRC07	aero6	inline	o	o	o	o	o	o	o	o	o
M16	CMAQ	5.0.2	SAPRC07	aero6	inline			o	o	o	o	o	su	su
M17	CMAQ	5.0.2	CB05	aero6	inline			o	o	o	o	o	su	su
M18	CMAQ	5.0.2	RACM2	aero6	inline			o	o	o	o	o	su	su
M19	CMAQ	5.0.2	SAPRC99	aero5	inline			o	o	o	o	o	su	su
M20	CMAQ	5.0.1	SAPRC99	aero5	inline	o	o	o		E2	D	W	o	
M21	CMAQ	5.0.1	SAPRC07	aero6	inline	o		o	o	o	o	o	o	o
M22	CMAQ	5.0.1	SAPRC07	aero6	inline			o	o	o	o	o	o	o
M23	CMAQ	5.0.1	SAPRC07	aero6	inline			o	o	o	o	o	o	o
M24	CMAQ	5.0.1	CB05	aero6	inline			o	o	o	o	o	o	o

Table 2. *Cont.*

ID	Model	Version	Chemical Mechanism	Aerosol Module	Photolysis	Simulation [1]				Emis [2]	BCON [3]	Met [4]	Submitted [5]	
						d01	d02	d03	d04				d03	d04
M25	CMAQ	5.0.1	SAPRC99	aero5	inline				o	o	o	o		o
M26	CMAQ	5.0.1	SAPRC99	aero5	inline			o	o	o	o	o	o	o
M27	CMAQ	4.7.1	SAPRC99	aero5	table				o	E3	o	o		o
M28	CMAQ	4.7.1	SAPRC99	aero5	table				o	o	o	o		o
M29	CAMx	6.4	SAPRC07	CF	table			o	o	o	o	o	o	o
M30 [6]	WRF-Chem	3.7.1	RADM2	MADE	inline	o	o	o	o	o	M	WC	o	o
M31	WRF-Chem	3.7.1	RADM2	MADE	inline	o	o	o	o	o	M	WC	su	su
M32	WRF-Chem	3.7.1	RADM2	MADE	inline	o	o	o	o	o	M	WC	su	su

[1] "o" indicates the domains that participants used to conduct their simulations. [2] Input emissions. "o" indicates that the baseline model-ready emission is used. "E1" uses EAGrid2010-JAPAN [48] and HTAP Version 2.2 [42]. "E2" uses EAGrid2000-JAPAN [49] and NASA INTEX-B [50]. "E3" uses EAGrid2000-JAPAN [49]. [3] Boundary concentration. "o" indicates that the baseline boundary concentration is used. "M" uses MOZART-4 [51]. "D" uses CMAQ defaults. [4] Meteorological condition. "o" indicates that the baseline metrological condition is used. "W" uses the meteorology simulated using WRF-ARW with own conditions, including physical options, parameterizations, and meteorological reanalysis. "WC" indicates the meteorology simulated using WRF-Chem with own conditions including physical options and parameterizations. [5] "o" indicates data submitted for d03 and d04 in each season. "su" was submitted for only summer. [6] NH_4^+ and total $PM_{2.5}$ were not submitted.

3. Observational Data for Model Evaluation

A monitoring framework of ambient $PM_{2.5}$ components was initiated in the fiscal year 2011 under the Japan government initiative [4]. Over a period of at least two weeks set for each season, 1-day accumulated concentrations of $PM_{2.5}$ components, including ions (e.g., SO_4^{2-}, NO_3^-, and NH_4^+), inorganic elements (e.g., Na, Al, K, and Ca), and carbonaceous aerosols (EC and OC), were monitored using the filter pack method at selected stations from three types of APMSs, including AAPMSs, roadside APMSs (RAPMSs), and background monitoring stations (BGMSs). $PM_{2.5}$ mass concentrations determined gravimetrically by weighing the filters were employed as the $PM_{2.5}$ mass concentration in this paper. Monitoring data from valid AAPMSs that obtained data for each $PM_{2.5}$ component over a period of at least eight days (53%) from each target period (up to 15 days) per each station were used to evaluate the performances of the participant CTMs. The number of valid AAPMSs was 16–22 stations for each domain and season. The data acquisition rate was highest in the summer, while a poor data acquisition rate was found for NO_3^- in autumn. Observed gaseous pollutants at these AAPMSs were also used to evaluate the simulated nitric oxide (NO), nitrogen dioxide (NO_2), and sulfur dioxide (SO_2).

Figures 4 and 5 present observed daily concentrations for $PM_{2.5}$ components, i.e., SO_4^{2-}, NO_3^-, NH_4^+, EC, and OC, and total $PM_{2.5}$ mass for each 12- to 15-day seasonal period at the AAPMSs for d03 and d04, respectively. The box-and-whisker and black dots (outliers) means the differences between AAPMSs in each domain.

In general, the concentrations of the total $PM_{2.5}$ and its components within a single domain exhibit similar day-to-day variabilities for each season. However, the frequency distributions of daily concentrations between the AAPMSs in each domain were enhanced, particularly for elevated concentrations (Figures 4 and 5). Therefore, the spatially averaged concentrations obtained from daily monitoring data for different AAPMSs within each domain were used for time series analysis hereafter.

For d03, i.e., western Japan, the seasonal-average total $PM_{2.5}$ concentrations were 17.3, 23.1, 20.3, and 19.6 $\mu g/m^3$, with maximum daily concentrations of 31.9, 37.6, 36.3, and 34.3 $\mu g/m^3$ for spring, summer, autumn, and winter. The summer $PM_{2.5}$ concentration was slightly higher than those for the other seasons; however, the seasonal characteristics of the $PM_{2.5}$ concentration are unclear. SO_4^{2-} was a dominant $PM_{2.5}$ component, accounting for approximately 40% (9.1 $\mu g/m^3$) of the total $PM_{2.5}$ mass concentration in the summer. Meanwhile, from autumn to winter, the ratios of NO_3^- and OC to total $PM_{2.5}$ mass increased. The ratios of the five major $PM_{2.5}$ components were similar, with values of 12%–19% in the winter. On the dates when $PM_{2.5}$ was elevated, the AAPMS differences in $PM_{2.5}$ concentration levels increased, and considerably high $PM_{2.5}$ was found at AAPMSs placed at major cities: Osaka and Nagoya. These results were compared with those from rural areas.

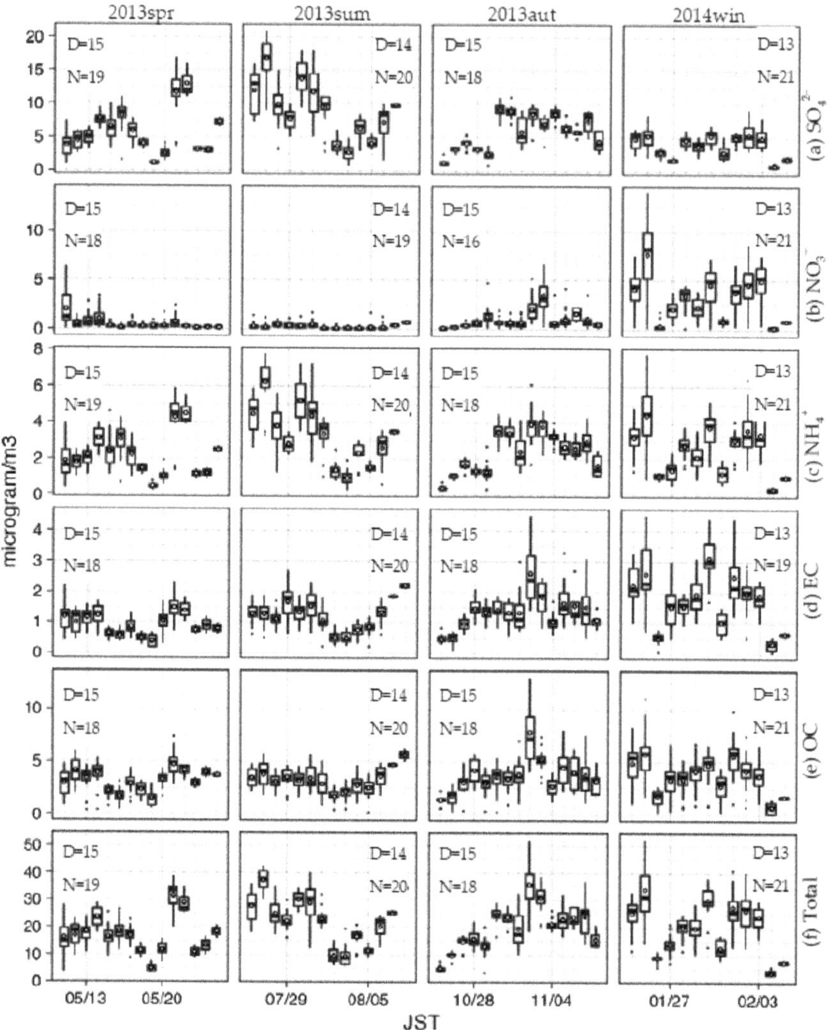

Figure 4. Box-plots of observed daily concentrations of total particulate matter with a diameter of 2.5 μm or less (PM$_{2.5}$) and its components: (**a**) sulfates (SO$_4^{2-}$), (**b**) nitrates (NO$_3^-$), (**c**) ammonium (NH$_4^+$), (**d**) elemental carbon (EC), (**e**) organic carbon (OC), and (**f**) total PM$_{2.5}$ mass, at AAPMSs within d03 for the four seasons. The open circles indicate spatially averages obtained from daily concentrations observed at AAPMSs within d03. The black dots indicate the outliers. The box-and-whisker and outliers represent the frequency distributions of daily concentrations observed at AAPMSs in d03. D presents the numbers of days with available observations, and N presents the number of AAPMSs.

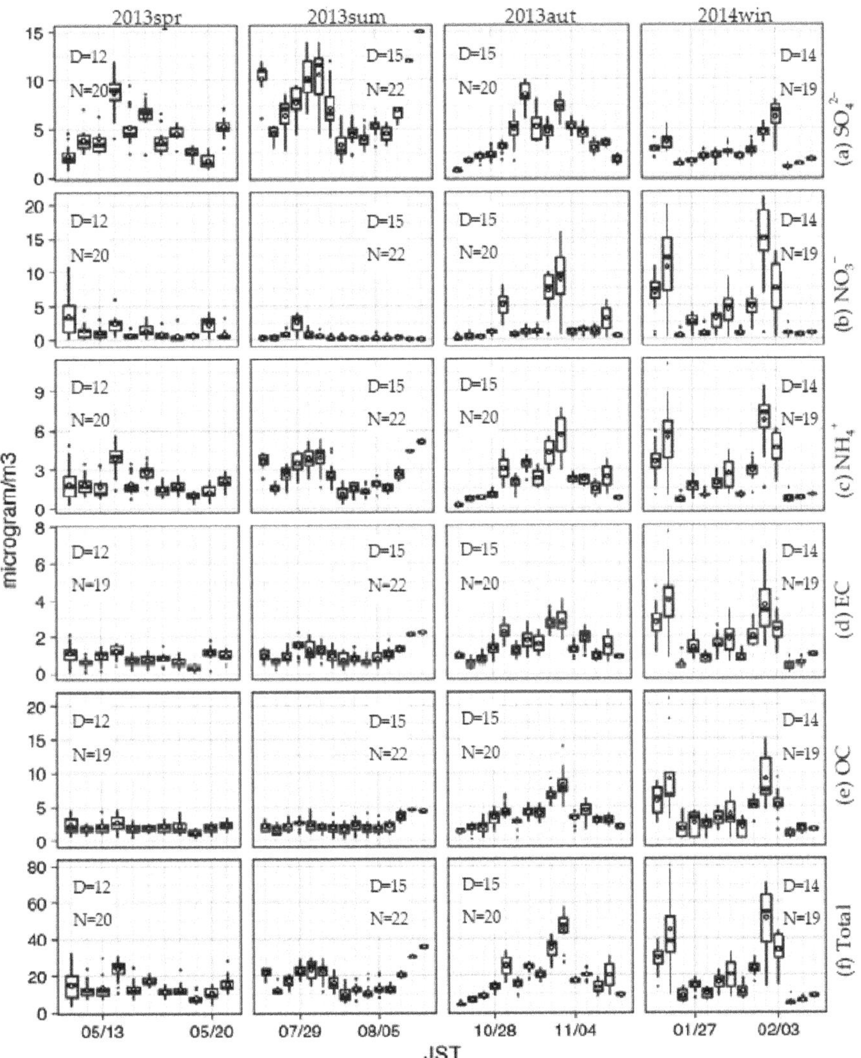

Figure 5. Box-plots of observed daily concentrations of total PM$_{2.5}$ and its components: (**a**) SO$_4^{2-}$, (**b**) NO$_3^-$, (**c**) NH$_4^+$, (**d**) EC, (**e**) OC, and (**f**) total PM$_{2.5}$ mass, at AAPMSs within d04 for the four seasons. The open circles indicate spatially averages obtained from daily concentrations observed at AAPMSs within d04. The black dots indicate the outliers. The box-and-whisker and outliers represent the frequency distributions of daily concentrations observed at AAPMSs in d04. D presents the numbers of days with available observations, and N presents the number of AAPMSs.

For d04, the Tokyo metropolitan area, which is several hundred kilometers east of d03, the day-to-day changes in the concentrations of total PM$_{2.5}$ and its components were similar to those for d03; however, the seasonal-average concentrations: 15.2, 18.6, 17.9, and 19.3 µg/m^3 for spring, summer, autumn, and winter, were slightly lower than those for d03; whereas the maximum daily concentrations were 26.3, 35.4, 41.9, and 46.8 µg/m^3. The elevated daily concentrations were obviously higher than those for d03 in the autumn and winter. Wintertime PM$_{2.5}$ concentrations were slightly higher than those in the other seasons, with increased daily concentrations; however, the seasonal characteristics of the $PM_{2.5}$ concentration were unclear for d04. The daily variabilities of the total $PM_{2.5}$ mass were characterized by SO_4^{2-} in spring and summer, where the ratios of SO_4^{2-} to total $PM_{2.5}$ mass were 32% (4.8 $\mu g/m^3$) and 39% (4.8 $\mu g/m^3$), respectively. In autumn, the ratios of the other $PM_{2.5}$ components, including OC, NO_3^-, and NH_4^+, to the total $PM_{2.5}$ mass increased. The OC and NO_3^- concentrations were both higher than the SO_4^{2-} concentration in winter. In particular, for the first pollutant peak on 25 January, OC and NO_3^- were dominant, accounting for 22% (9.2 $\mu g/m^3$) and 23% (9.6 $\mu g/m^3$) of the total $PM_{2.5}$ mass concentration, respectively. For the peak on 2 February, NO_3^- was dominant, accounting for 22% (9.2 $\mu g/m^3$). SO_4^{2-} was the dominant $PM_{2.5}$ component throughout the year for d03, but for d04, OC and NO_3^- levels were higher than SO_4^{2-} levels in the winter. On the dates $PM_{2.5}$ elevated, the AAPMSs differences of $PM_{2.5}$ concentration levels were increased, and the considerably high $PM_{2.5}$ were found at the AAPMSs placed on the central area of d04, i.e., the Tokyo metropolitan area.

4. Results and Discussion

4.1. Hourly Concentrations of Primary Pollutants

Major gaseous pollutants were also monitored at the AAPMSs. Figures 6 and 7 present the spatial averages of observed and simulated hourly concentrations of NO, NO_2, and SO_2 from different AAPMSs for d03 and d04, respectively. Table 3 summarizes the ensemble performances of the participant CTMs at each AAPMS for each season.

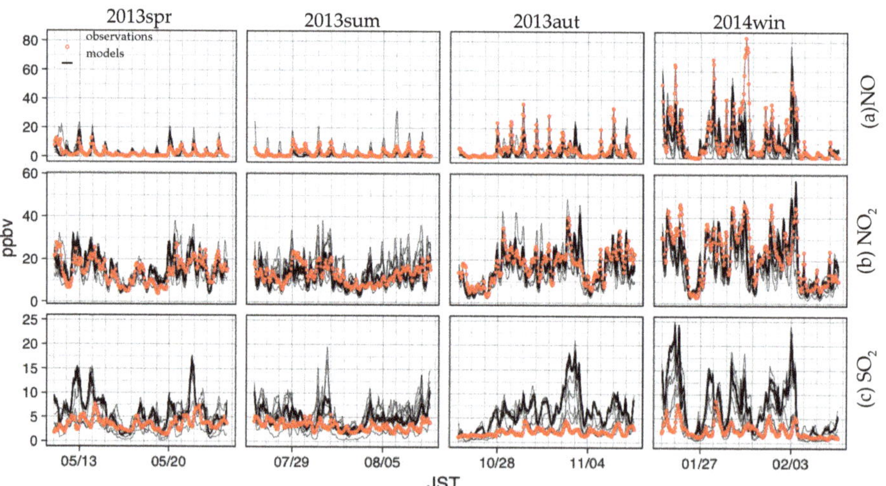

Figure 6. Spatial averages of observed and simulated hourly concentrations for (**a**) NO, (**b**) nitrogen dioxide (NO_2), and (**c**) sulfur oxide (SO_2) from different AAPMSs within d03 over the four seasons.

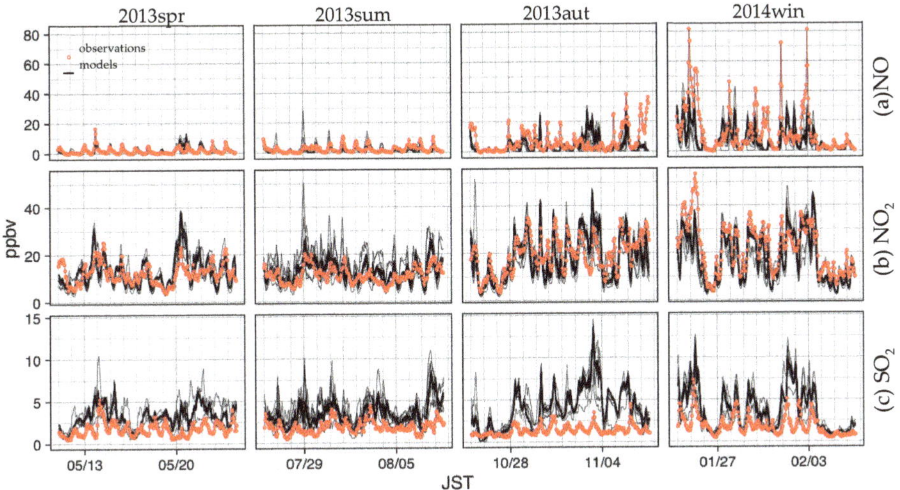

Figure 7. Spatial averages of observed and simulated hourly concentrations for (**a**) NO, (**b**) NO$_2$, and (**c**) SO$_2$ from different AAPMSs within d04 over the four seasons.

Table 3. Observed and simulated averaged concentrations [1] and ensemble performances [2] of the participant CTMs for hourly concentrations of NO, NO$_2$, and SO$_2$ at each AAPMS in each season.

	MEAN		NMB	Correl	IoA	N		MEAN		NMB	Correl	IoA	N
	[µg/m^3]		[%]					[µg/m^3]		[%]			
	Observation	Model						Observation	Model				
			d03							d04			
2013spr													
NO	2.53	2.53	8.3	0.38	0.48	24		2.53	2.53	58.26	0.37	0.50	22
NO$_2$	14.49	15.02	0.4	0.37	0.58	24		14.49	15.02	7.13	0.38	0.56	22
SO$_2$	3.61	5.70	120.0	0.33	0.42	17		3.61	5.70	128.23	0.33	0.46	20
2013sum													
NO	2.59	1.05	−53.0	0.43	0.50	24		5.10	3.00	−42.53	0.40	0.49	23
NO$_2$	12.75	12.31	−3.1	0.37	0.58	24		17.32	16.20	14.71	0.41	0.58	23
SO$_2$	3.28	4.78	189.0	0.19	0.35	16		2.18	5.98	119.09	0.28	0.40	20
2013aut													
NO	5.10	3.00	−36.2	0.34	0.47	24		2.59	1.05	−43.13	0.18	0.41	23
NO$_2$	17.32	16.20	−8.1	0.47	0.64	24		12.75	12.31	−4.94	0.45	0.64	23
SO$_2$	2.18	5.98	353.6	0.32	0.26	17		3.28	4.78	449.61	0.31	0.30	20
2014win													
NO	15.16	9.62	−41.5	0.31	0.51	24		15.16	9.62	−42.15	0.28	0.49	23
NO$_2$	22.55	19.50	−16.1	0.56	0.72	24		22.55	19.50	−19.40	0.54	0.71	23
SO$_2$	2.71	7.02	233.6	0.40	0.37	17		2.71	7.02	138.96	0.33	0.38	20

[1] Observation MEAN calculated from hourly value at each AAPMS in d03 and d04 for each season, respectively. Model MEAN calculated from seasonal averages from hourly value in each CTM corresponding to available observations at each AAPMS in d03 and d04. [2] Ensemble means of NMB (normalized mean bias), Correl (correlation coefficient), IoA (index of agreement), N (the number of available observation stations) calculated from all pairs of observations and simulations for each AAPMS and CTM in d03 and d04, respectively.

In general, most CTMs showed good agreement with the observed concentration levels of NO and NO$_2$ for each season, with regular diurnal patterns of NO in the warmer seasons. However, none of the models fully reproduced the elevated concentrations, e.g., for 19–20 May and 3 November (NO and NO$_2$) and 30 January (NO), with differences of 50%–200% between the observations and models, among others; the models tend to overestimate the observed daily maximums of NO: around 10–20 ppbv in spring and around 10–30 ppbv in summer, by a factor of 2.

For midnight on 30 January, all participant CTMs could not simulate the considerably increased level of NO before the rapid NO decrease associated with airmass changes, although all participants reproduced the NO decrease well. This suggests that CTMs successfully simulated the concentration change owing to meteorological changes in the synoptic scale but failed to simulate an increase in the amounts owing to local scale meteorological changes such as the strong atmospheric stability, especially during colder seasons. All models tended to overevaluate the daytime NO reduction. In particular, two WRF-Chem types (M30 and M31) and M05 produced strikingly low constant values, 0.001 or 0.000 ppbv, during the daylight hours in summer and autumn. The normalized mean bias (NMB) for both domains produced a strong underestimation of NO (approximately −40% to −50%), except during the spring. Underestimates of NO at remote stations in Japan have been observed for regional CTMs, as reported by MICS–Asia III results [9], and the correlations and index of agreement (IoA) values ranged from 0.18 to 0.43 and 0.41 to 0.51, respectively. The performance levels of each model exhibited substantial differences between both domains and seasons. The differences between seasons are likely related to meteorology simulation abilities, but the reasons for the differences appearing between domains are unclear in this stage.

The differences for NO_2 in each model were large. Among these models, M31, M32, and M30 tended to overestimate elevated NO_2 levels. The lower levels of NO_2 obtained by M30 were often comparable to the NO_2 concentration obtained by M03, which provided considerably lower NO_2 concentrations compared to other models. These results suggest that the differences in meteorological conditions and NO_x chemistry in each model produced the NO_2 discrepancy between the models. Most of the models produced better results for NO_2 than for NO, with ensemble averages of seasonal statistics, e.g., correlation values, of 0.56 (d03) and 0.55 (d04), 0.72 (d03), and 0.71 (d04), particularly in the winter.

Over the year, most models obviously overestimated the observed SO_2, with an ensemble bias of 1.7–4.2 ppbv (NMB: 120%–350%) for d03 and 1.5–2.5 ppbv (NMB: 160%–470%) for d04. In addition, relatively high SO_2 levels were found for M30, M31, and M32. Meanwhile, M03 and M20 tended to produce lower concentrations compared to the other models, with a negative bias of −1.3 ppbv (M03) and −0.2 ppbv (M20) recorded especially in the spring; and exhibited better performances (IoA: 0.58–0.59) over the other models (IoA: 0.30–0.39), especially in the winter. The input SO_2 emissions into two CMAQ simulations (M03 and M20) differed from SO_2 emissions of J-STREAM. For example, SO_2 emissions in both total and bottom layers of J-STREAM were more than twice those of M03 for d03, respectively. Meanwhile, for d04, including active volcanos, although the total SO_2 emissions of J-STREAM were half those of M03, the bottom layer SO_2 emissions of J-STREAM were 1.3 times those of M03. The differences in divided SO_2 emission amounts in the lower layers possibly affected the simulated atmospheric SO_2 concentrations. The second-best model setting, M03, performed slightly better (IoA: 0.41) than other models, which suggests that atmospheric SO_2 concentrations were considerably affected by the input emission conditions, including the injection heights. Although modifications of emission conditions help to produce better SO_2 simulation, using modifications alone to resolve the overestimation of SO_2 (up to 470%) is not realistic.

The differences among models with respect to emissions, chemistries, and meteorological conditions led to major differences in simulated primary pollutant concentrations; moreover, the simulated differences between similar model settings increased in the winter.

4.2. Simulated Daily Concentrations of $PM_{2.5}$ Components and Total $PM_{2.5}$ Mass

Figures 8 and 9 present spatially averages obtained from observed and simulated daily concentrations for $PM_{2.5}$ components (SO_4^{2-}, NO_3^-, NH_4^+, EC, and OC) and total $PM_{2.5}$ mass for different AAPMSs in d03 and d04, respectively. The seasonal ensemble performances of the participant CTMs at each AAPMS are also summarized as statistics in Tables 4 and 5 for each domain. The goal and criteria levels for CTM performance statistics, NMB, normalized mean error (NME), and correlation were recommended by Emery et al. [52], and the fractional bias (FB) and fractional error (FE) were recommended by Boylan and Russell [53], which is listed in Table A1. Individual model performance reports of each CTM are shown in Tables A2 and A3.

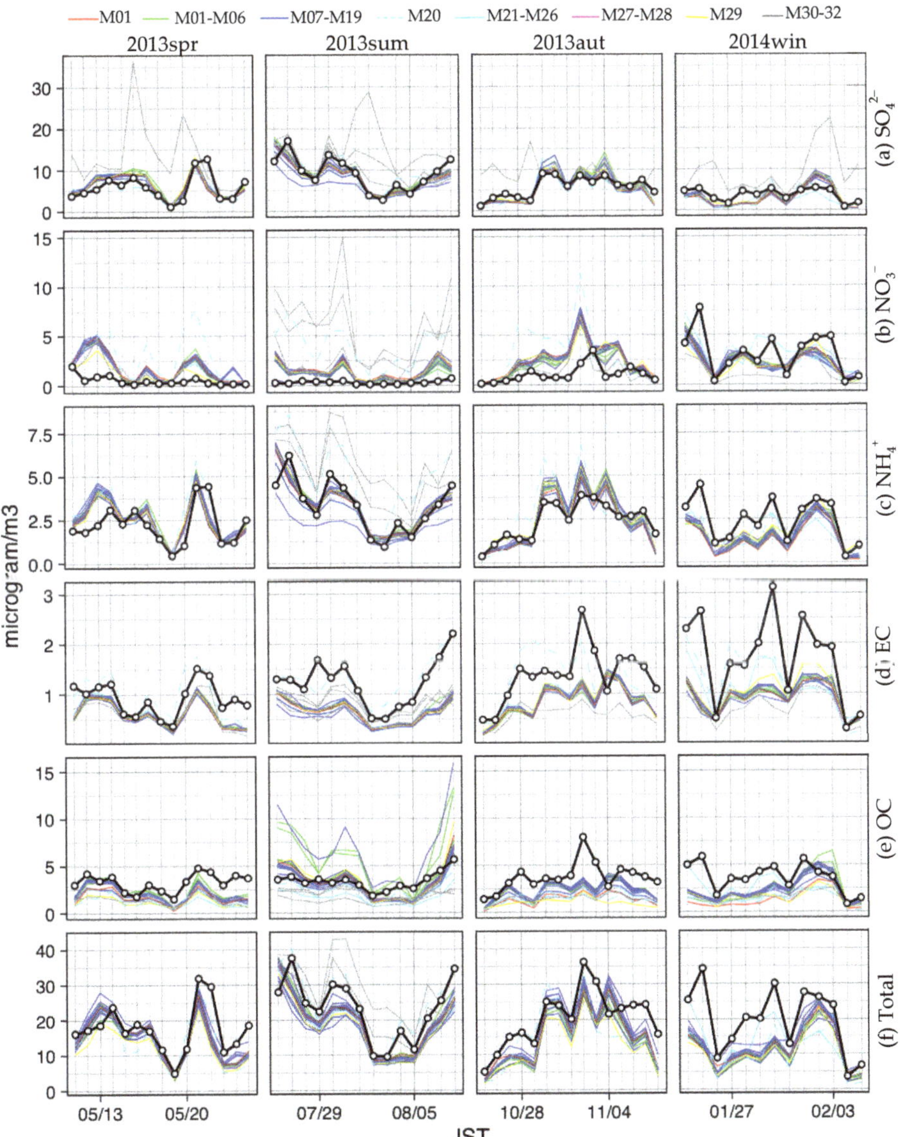

Figure 8. Spatially averaged concentrations of PM$_{2.5}$: (**a**) SO$_4^{2-}$, (**b**) NO$_3^-$, (**c**) NH$_4^+$, (**d**) EC, (**e**) OC, and (**f**) total PM$_{2.5}$ mass over the four seasons. These results are based on daily concentrations observed and simulated for AAPMSs in d03. The thick solid lines with open circles present observations, and the colored lines present model results.

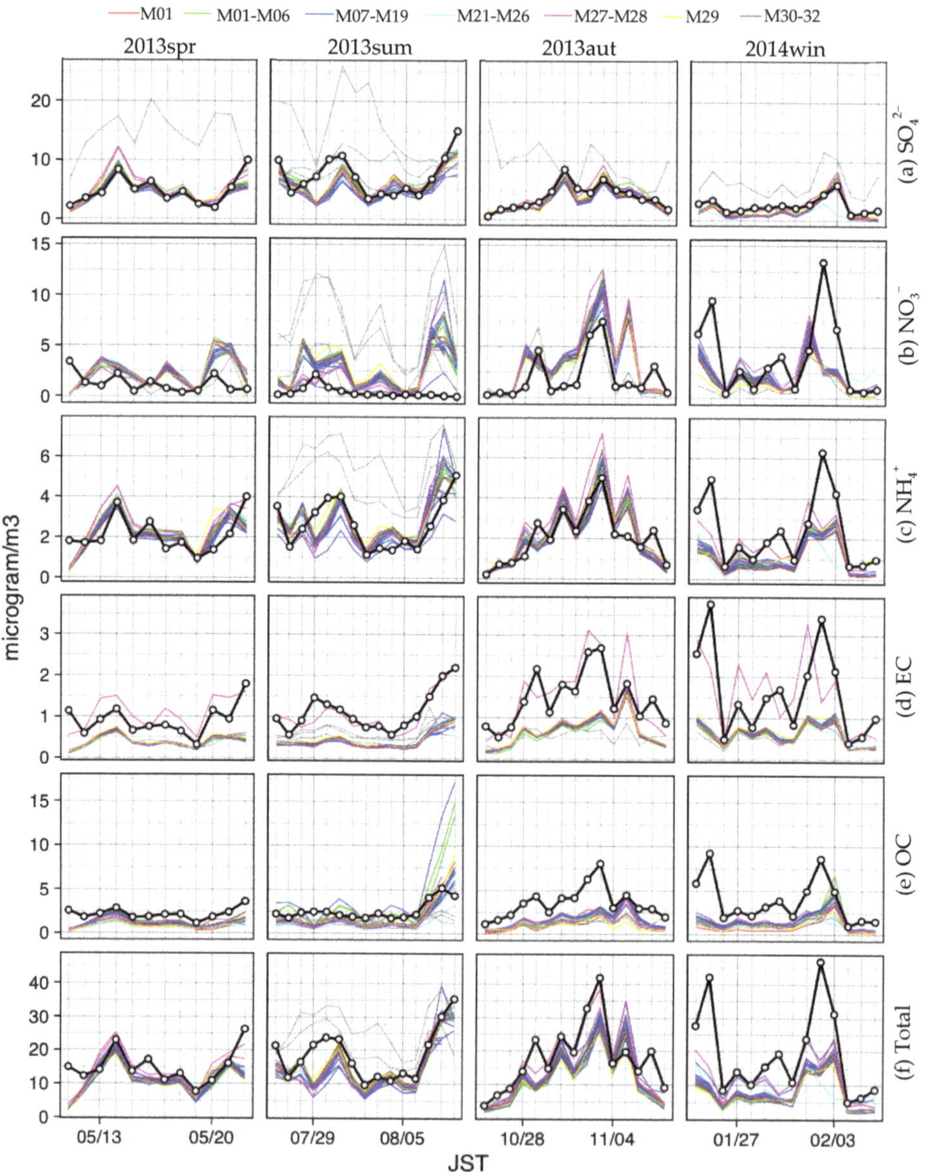

Figure 9. Spatially averaged concentrations of PM$_{2.5}$: (**a**) SO$_4{}^{2-}$, (**b**) NO$_3{}^-$, (**c**) NH$_4{}^+$, (**d**) EC, (**e**) OC, and (**f**) total PM$_{2.5}$ over the four seasons. These results are based on daily concentrations observed and simulated for AAPMSs in d04. The thick solid lines with open circles present observations, and the colored lines present model results.

Table 4. Observed and simulated averaged concentrations [1] and ensemble performances [2] of the participant CTMs for daily concentrations of total $PM_{2.5}$ and $PM_{2.5}$ components at each AAPMS within d03.

	MEAN		MB	ERROR	RMSE	NMB	NME	FB	FE	Correl	IoA	N
	[µg/m³]		[µg/m³]	[µg/m³]	[µg/m³]	[%]	[%]	[%]	[%]			
	Observation	Model										
2013spr												
SO_4^{2-}	5.87	6.59	1.68	2.23	2.95	32.44	41.39	25.75	34.46	0.77	0.79	19
NO_3^-	0.52	1.72	1.44	1.61	2.22	310.08	333.28	72.92	109.94	0.44	0.34	18
NH_4^+	2.21	2.54	0.65	0.96	1.21	32.01	45.82	26.86	39.52	0.70	0.74	19
EC	0.92	0.63	−0.20	0.31	0.39	−22.33	33.40	−27.91	38.94	0.73	0.71	18
OC	3.23	1.85	−1.06	1.21	1.42	−32.46	39.30	−45.81	52.95	0.70	0.67	18
TOTAL	17.33	14.58	−0.99	3.95	4.90	−5.30	22.60	−9.00	25.53	0.81	0.86	19
2013sum												
SO_4^{2-}	9.11	8.84	0.04	2.68	3.50	1.65	30.34	3.66	30.41	0.74	0.82	20
NO_3^-	0.28	1.73	1.22	1.30	1.73	650.97	672.36	98.26	122.69	0.32	0.21	19
NH_4^+	3.32	3.49	0.23	1.16	1.46	8.47	35.92	8.92	34.28	0.73	0.79	20
EC	1.23	0.69	−0.45	0.49	0.56	−38.53	42.31	−50.07	54.86	0.67	0.62	20
OC	3.35	3.28	−0.01	1.34	1.60	6.75	47.22	−11.01	45.68	0.65	0.58	20
TOTAL	23.07	19.71	−2.93	5.96	7.33	−12.77	26.30	−16.97	29.10	0.78	0.82	20
2013aut												
SO_4^{2-}	5.73	5.46	−0.21	1.73	2.12	0.50	29.11	−13.77	34.68	0.86	0.87	18
NO_3^-	1.05	2.33	1.55	1.96	2.70	193.57	233.81	54.87	103.17	0.28	0.31	16
NH_4^+	2.38	2.49	0.30	0.87	1.16	13.45	35.42	−2.69	34.45	0.86	0.83	18
EC	1.36	0.85	−0.50	0.65	0.78	−33.56	43.92	−44.22	55.76	0.43	0.56	18
OC	3.73	2.02	−1.78	2.03	2.43	−42.93	49.71	−57.62	67.42	0.44	0.51	18
TOTAL	20.27	15.36	−4.26	6.61	7.98	−19.16	30.78	−32.14	40.63	0.76	0.79	18
2014win												
SO_4^{2-}	3.75	3.31	−0.39	2.00	2.41	−8.85	52.14	−35.89	61.17	0.66	0.68	21
NO_3^-	3.08	2.31	−0.89	1.86	2.55	−18.85	57.79	−18.48	68.65	0.37	0.60	21
NH_4^+	2.40	1.51	−0.94	1.05	1.30	−34.93	39.89	−48.01	53.60	0.75	0.74	21
EC	1.69	0.94	−0.90	0.98	1.19	−45.98	50.26	−52.36	62.90	0.55	0.58	19
OC	3.62	2.07	−1.72	2.13	2.53	−35.59	57.36	−52.86	73.17	0.43	0.53	21
TOTAL	19.57	11.69	−8.66	9.12	11.21	−40.43	43.05	−52.25	55.77	0.67	0.65	21

[1] Observation MEAN calculated from daily value at each AAPMS in d03 for each season. Model MEAN calculated from seasonal averages from daily value in each CTM corresponding to available observations at each AAPMS in d03. [2] Ensemble mean of MB (mean bias), ERROR (mean error), RMSE (root mean square error), NMB (normalized mean bias), NME (normalized mean error), FB (fractional bias), FE (fractional error), Correl (correlation coefficient), IoA (index of agreement), N (the number of available observation stations) calculated from all pairs of observation and simulation for each AAPMS and CTM in d03. Observation data from valid AAPMSs that obtained data for each $PM_{2.5}$ component over a period of at least eight days (53%) from each target period (up to 15 days) per each station were used to evaluate the performances (MB, ERROR, RMSE, NMB, NME, FB, FE, Correl, and IoA) of the participant CTMs.

Table 5. Observed and simulated averaged concentrations [1] and ensemble performances [2] of the participant CTMs for daily concentrations of total $PM_{2.5}$ and $PM_{2.5}$ components at each AAPMS within d04.

	MEAN		MB	ERROR	RMSE	NMB	NME	FB	FE	Correl	IoA	N
	[µg/m³]		[µg/m³]	[µg/m³]	[µg/m³]	[%]	[%]	[%]	[%]			
	Observation	Model										
2013spr												
SO_4^{2-}	4.82	4.91	0.39	1.57	1.93	10.98	35.85	0.35	33.74	0.80	0.84	20
NO_3^-	1.24	2.20	0.95	1.82	2.46	145.14	206.47	35.06	107.72	0.17	0.35	20
NH_4^+	2.10	2.14	0.15	0.78	1.01	9.44	39.89	2.42	41.05	0.57	0.72	20
EC	0.91	0.43	−0.39	0.46	0.54	−40.16	57.18	−59.42	73.42	0.44	0.51	19
OC	2.19	1.12	−0.94	1.07	1.22	−39.05	50.83	−59.38	70.61	0.45	0.49	19
TOTAL	15.02	11.47	−2.91	4.59	6.11	−19.01	32.18	−27.57	39.98	0.53	0.68	20
2013sum												
SO_4^{2-}	6.49	6.25	−0.68	2.82	3.54	−9.21	44.02	−13.08	44.10	0.36	0.57	22
NO_3^-	0.40	2.85	1.85	1.92	2.68	587.86	600.84	117.21	129.97	0.44	0.34	22
NH_4^+	2.66	2.88	0.13	0.99	1.26	6.98	42.50	5.58	38.98	0.56	0.67	22
EC	1.13	0.47	−0.57	0.60	0.67	−55.79	59.89	−79.59	84.08	0.44	0.45	22
OC	2.56	1.93	−0.81	1.08	1.23	−32.37	49.92	−49.81	63.83	0.10	0.39	22
TOTAL	18.61	15.18	−3.35	5.97	7.36	−19.73	36.69	−25.14	41.25	0.52	0.61	22

Table 5. *Cont.*

	MEAN		MB	ERROR	RMSE	NMB	NME	FB	FE	Correl	IoA	N
	[μg/m^3]		[μg/m^3]	[μg/m^3]	[μg/m^3]	[%]	[%]	[%]	[%]			
	Observation	Model										
2013aut												
SO_4^{2-}	3.91	3.66	−0.38	1.30	1.63	−7.51	30.43	−14.04	32.83	0.84	0.86	20
NO_3^-	2.00	3.36	1.99	2.82	3.90	171.39	206.12	55.45	95.32	0.54	0.56	20
NH_4^+	2.07	2.15	0.24	0.91	1.24	14.50	42.85	5.85	39.32	0.73	0.78	20
EC	1.46	0.72	−0.79	0.84	1.01	−49.30	53.29	−67.93	71.71	0.51	0.57	20
OC	3.53	1.53	−2.19	2.27	2.70	−56.03	58.42	−79.11	84.80	0.55	0.53	20
TOTAL	17.97	13.07	−4.49	6.97	8.99	−22.54	35.93	−32.02	43.51	0.71	0.76	20
2014win												
SO_4^{2-}	2.55	1.92	−0.60	1.19	1.34	−20.85	42.55	−48.72	60.94	0.88	0.84	19
NO_3^-	3.91	2.06	−2.36	3.04	4.54	−42.80	65.21	−38.83	78.53	0.37	0.55	19
NH_4^+	2.31	1.10	−1.40	1.46	1.99	−49.28	51.85	−64.33	67.79	0.70	0.65	19
EC	1.61	0.67	−1.05	1.11	1.46	−57.57	61.09	−70.06	76.46	0.48	0.52	19
OC	3.82	1.41	−2.77	3.03	3.90	−58.50	68.46	−74.99	96.14	0.36	0.48	19
TOTAL	19.26	8.45	−12.53	12.76	16.88	−55.14	56.41	−71.88	73.82	0.62	0.58	19

[1] Observation MEAN calculated from daily value at each AAPMS in d04 for each season. Model MEAN calculated from seasonal averages from daily value in each CTM corresponding to available observations at each AAPMS in d04. [2] E Ensemble means of MB (mean bias), ERROR (mean error), RMSE (root mean square error), NMB (normalized mean bias), NME (normalized mean error), FB (fractional bias), FE (fractional error), Correl (correlation coefficient), IoA (index of agreement), N (the number of available observation stations) calculated from all pair of observation and simulation for each AAPMS and CTM in d04. Observation data from valid AAPMSs that obtained data for each PM$_{2.5}$ component over a period of at least eight days (53%) from each target period (up to 15 days) per each station were used to evaluate the performances (MB, ERROR, RMSE, NMB, NME, FB, FE, Correl, and IoA) of the participant CTMs.

With SO_4^{2-} as a dominant PM$_{2.5}$ component, most CTMs showed good agreement with daily concentration levels and day-to-day changes in both domains for each season, with the exception of a few model settings. Overall, the ensemble statistics, including the NMB (−0.85, 1.65%), NME (30.34, 29.11), FB (3.66, −13.77%), FE (30.41, 34.28), and correlation (0.74, 0.86), passed the goal level in d03 for summer and autumn. For d04, the NMB (−7.5%), NME (30.34), FB (−13.04%), FE (32.83%), and correlation (0.84) passed the goal level for summer. With the exception of d03 in winter and d04 in summer, the correlation and IoA indicated excellent performance, with maximum values of 0.74–0.88 and 0.79–0.87 for d04 in winter. Most CTMs underestimated the observed SO_4^{2-} in d04 on 29–30 July, with relatively low values for the correlation (0.36) and IoA (0.52). This result may lead to underestimations of the total PM$_{2.5}$ mass in connection with the NH_4^+ concentrations. WRF-Chem (M30, 31) clearly overestimated SO_4^{2-} concentrations in PM$_{2.5}$ due to the SO_4^{2-} mass build-up problem associated with the nucleation calculation in MADE/SORGAM [54]. In addition, the WRF-Chem group employed their own physical parameterizations such as cumulus convection and microphysics for their meteorological simulations. Additional sensitivity simulations for meteorological fields are required to quantitatively evaluate the model inter-differences of SO_4^{2-} and total PM$_{2.5}$ mass concentrations owing to the differences in meteorological simulations. We will perform this in the next phase. The largest positive biases were found in M31, with 3.0–9.7 μg/m^3 (NMB: 52%–177%) for d03 and 3.8–10.2 μg/m^3 (NMB: 131%–240%) for d04. These simulated overestimations were slightly higher for CMAQ Version 4.7.1 (M27 and M28), particularly for d04 in spring. This trend indicates that the updated sulfur chemistries in CMAQ Version 5.0 [35,55–58] enhanced the performance of this model compared to the previous versions. In winter, CAMx (M29) performed better, with biases of −0.32 μg/m^3 (NMB: −7.3%) for d04 and 0.39 μg/m^3 (NMB: 3.4%) for d04 under the same emission condition. This result is attributed to an underestimation of SO_4^{2-} by the dominant participant model, CMAQ, which may be caused by an inadequate aqueous-phase SO_4^{2-} production by Fe- and Mn-catalyzed O_2 oxidation [14].

All participant CTMs overestimated NO_3^- levels in warmer seasons, with ensemble biases of 1.22–1.55 μg/m^3 (NMB: 194%–651%) for d03 and 0.85–1.99 μg/m^3 (NMB: 145%–588%) for d04. The largest positive biases were found in summer. Above all, M20, M30, and M31 strongly overestimated elevated NO_3^- levels. Only M11 showed relatively good agreement with observations for d03 in summer, with a minimum bias of 0.12 μg/m^3 (NMB: 91%) and improved values for the correlation (0.46) and IoA (0.54). However, M11 also produced low concentrations for $SO_4{}^{2-}$ and $NH_4{}^+$. As observed for d04 in autumn, all models exhibited better performance for the daily concentration levels and day-to-day changes in $NO_3{}^-$. For example, M30 has a minimum bias of 0.14 µg/m^3 (NMB: 11%), which passed the goal NMB level for 24-h $NO_3{}^-$. Some deviations in $NO_3{}^-$ between observations and the models were attributed to $NH_4{}^+$ and potentially NH_4NO_3. In winter, most models reproduced day-to-day changes in both domains but tended to underestimate elevated $NO_3{}^-$ levels, with ensemble mean biases of −0.89 µg/m^3 (NMB: −18.9%) and −2.36 µg/m^3 (NMB: −42.8%). A previous model inter-comparison study for the Tokyo metropolitan area, UMICS, concluded that the participant models overestimated $NO_3{}^-$ levels in both summer and winter [11,12], although available observations included only one winter and three summer stations. In our validations, most models produced higher $NO_3{}^-$ levels in spring and summer, lower $NO_3{}^-$ levels in winter, and moderate $NO_3{}^-$ levels in autumn, compared with accumulated observation data for d03 and d04. This result is expected to be more accurate than previous reports because a greater number of observations (for 18–22 stations) were included.

As mentioned above, the day-to-day variations in $NH_4{}^+$ were consistent with those of $SO_4{}^{2-}$ and $NO_3{}^-$. Therefore, most CTMs showed good agreement with daily concentration levels and day-to-day changes in both domains for each season, with the exception of some elevated peaks. Above all, the ensemble performances indicators, FE and FB, were −27.9%–8.9% and 34.3%–41.1%, thus passing the goal level in both domains for all seasons except winter. Notably, the differences among models increased in summer. Two WRF-Chem models (M32, M31) predicted higher $NH_4{}^+$ levels, with biases of 1.96–3.03 µg/m^3 (NMB: 84%–130%) and 1.72–2.71 µg/m^3 (NMB: 85%–61%) for d03 and d04, respectively. The M20 model, which employed EAGrid for emissions and an original configuration for meteorology, also produced relatively high $NH_4{}^+$ levels in d03, with a bias of 1.68 µg/m^3 (NMB: 51%). These overpredictions were likely associated with those of $SO_4{}^{2-}$ and $NO_3{}^-$ in summer. Meanwhile, relatively larger negative biases were found for M11, at −1.18 µg/m^3 (NMB: 35%) for d03 and −0.81 µg/m^3 (NMB: 33%) for d04.

The EC levels simulated by most CTMs were considerably lower than the observations in both domains for all season. The model ensemble biases were −0.90 to −0.20 µg/m^3 (NMB: −46% to −22%) and −2.77 to −0.39 µg/m^3 (NMB: −58% to −40%) for d03 and d04, respectively, with larger values for Tokyo. Both models employing EAGrid2000-JAPAN (M20 (d03) and M27 (d04)) produced higher EC values than other CTMs with different emission settings, and relatively better NMB values were obtained, at −20% to −3% and −35% to 42%, respectively. This trend suggests that the EC emissions of J-STREAM might be underestimated.

The CTMs reproduced some of elevated OC levels in the warmer seasons, but clearly underestimated the observed OC levels for autumn and winter, with model ensemble biases of −1.78 to −0.01 µg/m^3 (NMB: −42% to 7%) and −2.77 to −0.81 µg/m^3 (NMB: −59% to −39%) for d03 and d04, respectively, which are similar to the EC values. Additionally, as observed for the EC, the negative biases of OC for the Tokyo area were larger than those for western Japan. However, the negative biases of all participant CTMs have been clearly moderated compared with the UMICS cases [11,12]. Among the models, M02, M03, and M11 predicted relatively higher OC levels and overestimated the summer OC concentrations. Full-domain nesting simulations were performed via M02 and M03 using a relatively recent CMAQ model (Version 5.1), which includes updates for some chemical and aerosol mechanisms, such as POA aging, SOA mass yields with new pathways from isoprene, alkanes, and PAHs, and SOA formation reactions in the aqueous-phase chemistry. Continual nesting simulations for the Asian scale (d01) performed by CMAQ Version 5.1 exhibited higher regional-scale OC levels, leading to higher OC levels in urban areas in Japan compared with previous versions. Thus, an empirical SOA yield model can predict the same OC concentration level as the VBS model M11. It should be noted that effect of the updated SOA yield mechanisms was not clear at the urban scale when using CMAQ Version 5.1 or higher (e.g., M01, M04–05). Additionally, to evaluate simulated OC concentrations, more observational data are needed.

Overall, most CTMs showed good agreement with observed concentration levels of total $PM_{2.5}$ mass in both domains for each season. These results are likely associated with the reproducibility of some dominant components, e.g., SO_4^{2-} and NH_4^+. Moreover, CTMs tended to fail at reproducing some heavily polluted situations and underestimated the considerably high $PM_{2.5}$ concentrations (approximately 40–50 $\mu g/m^3$). A considerable underestimation ($\approx$30 $\mu g/m^3$) of total $PM_{2.5}$ associated with $PM_{2.5}$ components, except for SO_4^{2-}, was observed for d04 in the winter season, 25 January and 2 February; during that time, the nighttime simulated surface temperature was clearly lower than that in the observations (Figure 3). This implies that the simulated higher surface temperature compared with that in the observations formed weaker atmospheric stability, which produced weaker accumulations of particulate pollutants at nighttime, especially during colder seasons. The model ensemble biases were −8.66 to −0.99 $\mu g/m^3$ (NMB: −43% to −5%) for d03 and −2.91 to −11.98 $\mu g/m^3$ (NMB: −55% to −19%) for d04. The largest negative biases are found in winter due to underestimations of NH_4NO_3, particularly for d04. M31 and M32 tended to overpredict the total $PM_{2.5}$ due to overestimates of inorganic compounds. Of the model ensemble statistics for d03, the NMB (−5%, 13%) NME (22%, 26%), FB (−9%, −17%), FE (26%, 29%), and correlation (0.81, 0.78) passed the goal level for 24-h total $PM_{2.5}$ mass in spring and summer, respectively. In addition, the majority of the other statistical indicators passed the criteria levels as well.

5. Summery

A model inter-comparison of secondary pollutant simulations over urban areas in Japan, J-STREAM Phase I, was performed, in which a total of 32 simulations were conducted by combining CMAQ, CAMx, and WRF-Chem.

Simulated hourly concentrations of the primary pollutants NO and NO_2, which are precursors of $PM_{2.5}$, generally showed good agreement with the observed concentrations, at the same level as the MICS case. However, some differences between observations and simulations and CTMs may be considered to be caused by the differences in meteorological conditions and NOx chemistries of each CTM. Furthermore, most of the CTMs using the same input emissions tended to overestimate SO_2 concentrations, although the models showed good performance for $PM_{2.5}$ SO_4^{2-}. The different emission inventory, EAGrid produced better results for SO_2; therefore, it appears that the emission input can be improved. However, it was likely to be unrealistic that just the modifications of the emissions could fully resolve the overestimation of SO_2.

Simulated concentrations of $PM_{2.5}$ and its components were evaluated via a comparison with daily observed concentrations by using the filter pack method at selected AAPMSs for a period of at least two weeks for each season in this project. In general, most of the models showed good agreement with the observed concentration of total $PM_{2.5}$ mass for each season, within goal or criteria levels of model ensemble statistics especially in warmer seasons. This agreement was associated with the reproducibility of some of dominant particulates.

Among individual $PM_{2.5}$ components, most model results for SO_4^{2-} and NH_4^+ showed good agreement with daily concentration levels and day-to-day variations, with good model ensemble statistics, particularly for the warmer seasons. However, for SO_4^{2-}, a problem in the WRF-Chem model and novel, improved mechanisms for SO_4^{2-} formation in most CTMs were found through this model inter-comparison. Additionally, we found that the differences in the Asian scale precipitation patterns between precipitation parametrizations affected the simulated water-soluble $PM_{2.5}$ concentrations. Additional improvements for SO_4^{2-} were expected, particularly for the winter [11]. All participant models showed a strong tendency to overestimate NO_3^- in warmer seasons, with the model ensemble NMB reaching 651%. However, in winter, most of the models reproduced the day-to-day variations, with underestimations for elevated NO_3^- levels. These tendencies differed from a previous model inter-comparison, UMICS, which concluded that the participant models overestimated NO_3^- levels in both summer and winter [11,12]. This difference between two model inter-comparison studies is attributed to variations in the number of observations applied for verification. Thus, a sufficient amount of observation data on $PM_{2.5}$ components is needed to evaluate and improve CTMs. The EC levels simulated by most models were considerably lower than the observed levels for all seasons; however, some models employing EAGrid emissions produced higher EC levels than the other models. The models reproduced concentrations for some elevated OC values in the warmer seasons, but clearly underestimated the OC levels in autumn and winter. In addition, some models employing the VBS model and the newly updated SOA yield mechanisms produced higher OC levels and even overestimated the observed OC concentration in some cases.

This study has identified some effective approaches for improving $PM_{2.5}$ simulations for urban areas in Japan based on a model inter-comparison. First, improvements in emissions are expected to increase the reproducibility of primary pollutants that are precursors of $PM_{2.5}$ and EC concentrations. For SO_4^{2-}, NO_3^-, and OC, additional formation pathways can help to reduce underestimations. The recent model updates (e.g., CMAQ Version 5.3) improved the chemical pathways and are expected to simulate the secondary $PM_{2.5}$ components well. Simulated meteorological fields will be important for the Asian scale $PM_{2.5}$ concentration levels and the elevated $PM_{2.5}$ concentrations during the days with high amounts of pollution. In addition, special attention is needed for misjudgments in these models. Finally, additional accumulated observations are needed to evaluate the simulated concentrations. Future studies will include these modifications to realize reference air quality modeling in the next stages of J-STREAM.

Author Contributions: K.Y. managed the model inter-comparison (J-STREAM), prepared the initial, boundary, and meteorological inputs, and wrote this article. S.C. is the leader of J-STREAM project and prepared emission inputs. S.I. and H.H. are core members of J-STREAM in charge of inorganic aerosols. I.S. is a core member of J-STREAM in charge of photochemical gases. M.S., M.T., T.M. (Tazuko Morikawa), I.K., Y.M. (Yukako Miya), H.K., Y.M. (Yu Morino), K.K., T.N., H.S., K.U., and Y.F. are participants who conducted the model simulations. T.H. and T.M. (Takeshi Misaki) analyzed the submitted data. K.S. is in charge of global simulations. All authors have read and agreed to the published version of the manuscript.

Funding: This research was funded by the Environment Research and Technology Development Fund (5-1601) of the Environmental Restoration and Conservation Agency, JSPS KAKENHI Grant Number 18H03369, and the Collaborative Research Program of Research Institute for Applied Mechanics, Kyushu University.

Acknowledgments: This project was supported by the Environment Research and Technology Development Fund (5-1601) of the Environmental Restoration and Conservation Agency. This work was also partially supported by JSPS KAKENHI Grant Number 18H03369. Monitoring data of APMSs were obtained from National Institute for Environmental Studies. This work was supported in part by the Collaborative Research Program of Research Institute for Applied Mechanics, Kyushu University.

Conflicts of Interest: The authors declare no conflict of interest.

Appendix A

Table A1. Recommended benchmarks for photochemical model performance statistics [52,53].

Species	NMB		NME		r		FB		FE	
	Goal	Criteria	Goal	Criteria	Goal	Criteria	Goal	Criteria	Goal	Criteria
24-h $PM_{2.5}$, SO_4^{2-}, NH_4^+	<±10%	<±30%	<35%	<50%	>0.70	>0.40				
24-h NO_3^-	<±15%	<±65%	<65%	<115%	None	None	<±30%	<±60%	<±60%	<±75%
24-h OC	<±15%	<±50%	<45%	<65%	None	None				
24-h EC	<±20%	<±40%	<50%	<75%	None	None				

Table A2. Individual model performance (IoA) for d03.

ID	2013Spr						2013Sum					
	SO_4^{2-}	NO_3^-	NH_4^+	EC	OC	TOTAL	SO_4^{2-}	NO_3^-	NH_4^+	EC	OC	TOTAL
M01	0.82	0.32	0.74	0.71	0.62	0.86	0.84	0.19	0.80	0.60	0.62	0.83
M02	0.80	0.35	0.74	0.73	0.76	0.87	0.84	0.24	0.81	0.62	0.39	0.84
M03	0.81	0.34	0.75	0.74	0.70	0.88	0.85	0.23	0.82	0.59	0.34	0.83
M04	0.81	0.36	0.74	0.71	0.73	0.86	0.83	0.23	0.80	0.61	0.64	0.83
M05	0.80	0.33	0.72	0.72	0.74	0.86	0.84	0.22	0.80	0.62	0.63	0.83
M06	0.85	0.35	0.76	0.69	0.71	0.85	0.82	0.25	0.80	0.57	0.65	0.78
M07	0.82	0.35	0.75	0.71	0.70	0.85	0.87	0.23	0.83	0.65	0.61	0.86
M08	0.82	0.33	0.74	0.73	0.72	0.86	0.86	0.20	0.81	0.63	0.64	0.83
M09	0.83	0.33	0.75	0.73	0.72	0.86	0.84	0.20	0.81	0.63	0.64	0.82
M10							0.84	0.20	0.81	0.62	0.64	0.82
M11							0.69	0.43	0.66	0.54	0.30	0.76
M12	0.80	0.32	0.72	0.73	0.73	0.86	0.86	0.19	0.81	0.62	0.63	0.84
M13	0.81	0.34	0.70	0.73	0.62	0.86	0.85	0.19	0.80	0.63	0.64	0.83
M14	0.83	0.33	0.74	0.72	0.72	0.85	0.86	0.21	0.82	0.63	0.64	0.79
M15	0.83	0.33	0.74	0.72	0.71	0.87	0.86	0.21	0.82	0.62	0.64	0.84
M16							0.85	0.21	0.81	0.62	0.64	0.82
M17							0.85	0.20	0.81	0.62	0.64	0.83
M18							0.85	0.20	0.81	0.62	0.64	0.83
M19							0.85	0.20	0.80	0.62	0.63	0.83
M20	0.84	0.20	0.69	0.70	0.55	0.81	0.85	0.07	0.71	0.63	0.59	0.84
M21	0.82	0.32	0.73	0.72	0.72	0.85	0.85	0.19	0.81	0.63	0.64	0.83
M22	0.82	0.32	0.74	0.73	0.59	0.86	0.85	0.21	0.81	0.62	0.62	0.82
M23	0.84	0.36	0.76	0.71	0.70	0.87	0.86	0.22	0.82	0.62	0.64	0.83
M24	0.84	0.36	0.76	0.71	0.70	0.87	0.86	0.22	0.82	0.62	0.64	0.83
M26	0.83	0.37	0.73	0.71	0.50	0.87	0.86	0.21	0.81	0.62	0.50	0.81
M29	0.81	0.42	0.75	0.71	0.54	0.80	0.86	0.22	0.81	0.62	0.59	0.81
M30	0.27	0.32		0.66	0.50		0.36	0.22	0.57	0.54	0.42	0.69
M31							0.68	0.06	0.70	0.69	0.54	0.83
M32							0.84	0.07		0.68	0.52	

ID	2013Aut						2013Win					
	SO_4^{2-}	NO_3^-	NH_4^+	EC	OC	TOTAL	SO_4^{2-}	NO_3^-	NH_4^+	EC	OC	TOTAL
M01	0.91	0.30	0.85	0.85	0.46	0.79	0.69	0.60	0.69	0.55	0.45	0.61
M02	0.89	0.35	0.83	0.83	0.56	0.77	0.70	0.57	0.74	0.57	0.53	0.63
M03	0.84	0.38	0.80	0.80	0.50	0.77	0.70	0.57	0.76	0.57	0.47	0.63
M04	0.90	0.34	0.85	0.85	0.53	0.79	0.69	0.57	0.71	0.55	0.53	0.64
M05	0.90	0.32	0.84	0.84	0.54	0.80	0.69	0.58	0.72	0.56	0.53	0.65
M06	0.90	0.32	0.85	0.85	0.51	0.77	0.70	0.58	0.68	0.55	0.53	0.62
M07	0.91	0.29	0.84	0.84	0.52	0.80	0.69	0.51	0.69	0.52	0.51	0.64
M08	0.90	0.30	0.83	0.83	0.54	0.80	0.71	0.63	0.75	0.58	0.55	0.67
M09	0.90	0.30	0.83	0.83	0.54	0.80	0.69	0.61	0.74	0.58	0.55	0.67
M10												
M11							0.42	0.26	0.39	0.44	0.31	0.40
M12	0.89	0.29	0.80	0.80	0.55	0.81	0.71	0.60	0.76	0.58	0.56	0.69
M13	0.86	0.33	0.79	0.79	0.49	0.81	0.71	0.58	0.77	0.58	0.53	0.67
M14	0.91	0.31	0.83	0.83	0.55	0.76	0.70	0.61	0.74	0.58	0.54	0.62
M15	0.91	0.30	0.84	0.84	0.54	0.80	0.70	0.61	0.74	0.57	0.54	0.66
M16												
M17												
M18												
M19												
M20	0.88	0.19	0.70	0.70	0.49	0.74	0.71	0.56	0.78	0.65	0.50	0.67
M21	0.91	0.29	0.83	0.83	0.56	0.80	0.68	0.57	0.69	0.63	0.60	0.65
M22	0.90	0.30	0.83	0.83	0.48	0.80	0.46	0.03	0.34	0.44	0.35	0.35

Table A2. *Cont.*

ID	2013Aut						2013Win					
	SO_4^{2-}	NO_3^-	NH_4^+	EC	OC	TOTAL	SO_4^{2-}	NO_3^-	NH_4^+	EC	OC	TOTAL
M23	0.90	0.33	0.85	0.85	0.53	0.79	0.69	0.60	0.71	0.57	0.54	0.64
M24	0.90	0.34	0.85	0.85	0.53	0.79	0.69	0.60	0.71	0.57	0.54	0.64
M26	0.90	0.33	0.84	0.84	0.44	0.79	0.70	0.58	0.73	0.57	0.49	0.64
M29	0.91	0.32	0.83	0.83	0.44	0.74	0.70	0.62	0.72	0.60	0.50	0.61
M30	0.30	0.36			0.44		0.27	0.52		0.53	0.48	
M31							0.72	0.19	0.31	0.41	0.33	0.46
M32							0.42	0.11	0.31	0.40	0.32	0.36
M31												
M32												

Table A3. Individual model performance (IoA) for d04.

ID	2013Spr						2013Sum					
	SO_4^{2-}	NO_3^-	NH_4^+	EC	OC	TOTAL	SO_4^{2-}	NO_3^-	NH_4^+	EC	OC	TOTAL
M01	0.86	0.34	0.71	0.50	0.48	0.67	0.59	0.33	0.71	0.44	0.38	0.64
M02	0.88	0.38	0.76	0.52	0.53	0.69	0.59	0.34	0.71	0.44	0.29	0.63
M03	0.86	0.39	0.74	0.51	0.53	0.70	0.61	0.38	0.70	0.43	0.27	0.62
M04	0.87	0.35	0.72	0.51	0.51	0.69	0.63	0.32	0.73	0.44	0.41	0.67
M05	0.87	0.35	0.72	0.51	0.50	0.68	0.62	0.33	0.72	0.44	0.41	0.66
M06	0.86	0.36	0.71	0.50	0.48	0.64	0.56	0.40	0.68	0.43	0.41	0.59
M07	0.86	0.38	0.72	0.51	0.50	0.70	0.62	0.37	0.71	0.43	0.38	0.61
M08	0.87	0.35	0.71	0.51	0.51	0.67	0.56	0.36	0.66	0.44	0.39	0.60
M09	0.87	0.37	0.73	0.51	0.51	0.67	0.56	0.37	0.66	0.44	0.39	0.60
M10							0.57	0.35	0.67	0.44	0.41	0.61
M11							0.52	0.54	0.56	0.44	0.28	0.52
M12	0.88	0.35	0.71	0.51	0.52	0.68	0.56	0.34	0.66	0.44	0.39	0.60
M13	0.88	0.36	0.70	0.51	0.47	0.67	0.55	0.34	0.68	0.44	0.39	0.62
M14	0.88	0.34	0.73	0.51	0.51	0.67	0.56	0.34	0.66	0.44	0.40	0.58
M15	0.88	0.34	0.73	0.51	0.51	0.70	0.55	0.34	0.66	0.44	0.39	0.59
M16							0.57	0.34	0.67	0.44	0.41	0.61
M17							0.57	0.35	0.67	0.44	0.41	0.61
M18							0.58	0.33	0.68	0.44	0.41	0.62
M19							0.57	0.33	0.69	0.44	0.40	0.63
M21	0.88	0.34	0.72	0.51	0.50	0.68	0.57	0.35	0.67	0.44	0.41	0.61
M22	0.88	0.34	0.72	0.51	0.45	0.68	0.57	0.35	0.67	0.44	0.40	0.62
M23	0.88	0.36	0.74	0.51	0.51	0.70	0.58	0.33	0.69	0.43	0.40	0.62
M24	0.88	0.36	0.75	0.51	0.51	0.70	0.58	0.34	0.69	0.43	0.40	0.62
M25	0.89	0.36	0.73	0.51	0.46	0.71	0.55	0.37	0.69	0.42	0.38	0.61
M26	0.88	0.35	0.72	0.51	0.42	0.69	0.57	0.32	0.70	0.43	0.36	0.63
M27	0.79	0.38	0.67	0.54	0.53	0.68	0.60	0.41	0.71	0.56	0.44	0.63
M28	0.82	0.34	0.66	0.52	0.47	0.68	0.58	0.36	0.70	0.43	0.37	0.59
M29	0.86	0.36	0.66	0.52	0.43	0.64	0.59	0.31	0.73	0.46	0.42	0.67
M30	0.27	0.31		0.47	0.46		0.32	0.37		0.43	0.39	
M31							0.43	0.15	0.40	0.55	0.52	0.52
M32							0.65	0.16	0.55	0.53	0.54	0.70

ID	2013Aut						2013Win					
	SO_4^{2-}	NO_3^-	NH_4^+	EC	OC	TOTAL	SO_4^{2-}	NO_3^-	NH_4^+	EC	OC	TOTAL
M01	0.88	0.59	0.82	0.56	0.76	0.76	0.86	0.57	0.64	0.51	0.46	0.56
M02	0.91	0.58	0.80	0.57	0.76	0.76	0.88	0.56	0.67	0.52	0.50	0.57
M03	0.90	0.59	0.78	0.54	0.76	0.77	0.88	0.58	0.68	0.50	0.47	0.57
M04	0.89	0.58	0.80	0.56	0.77	0.77	0.87	0.55	0.65	0.51	0.49	0.58
M05	0.89	0.58	0.80	0.57	0.79	0.79	0.88	0.57	0.66	0.52	0.49	0.58
M06	0.85	0.58	0.80	0.56	0.73	0.73	0.86	0.56	0.63	0.51	0.49	0.57

Table A3. *Cont.*

ID	2013Aut						2013Win					
	SO_4^{2-}	NO_3^-	NH_4^+	EC	OC	TOTAL	SO_4^{2-}	NO_3^-	NH_4^+	EC	OC	TOTAL
M07	0.88	0.54	0.77	0.57	0.77	0.77	0.90	0.56	0.67	0.52	0.50	0.63
M08	0.88	0.55	0.78	0.57	0.76	0.76	0.88	0.57	0.66	0.52	0.50	0.59
M09	0.88	0.55	0.78	0.57	0.75	0.75	0.87	0.56	0.66	0.52	0.50	0.59
M10												
M11												
M12	0.90	0.54	0.77	0.57	0.77	0.77	0.89	0.57	0.68	0.52	0.50	0.60
M13	0.91	0.57	0.77	0.56	0.77	0.77	0.89	0.55	0.69	0.52	0.47	0.59
M14	0.89	0.55	0.78	0.57	0.73	0.73	0.87	0.58	0.67	0.53	0.50	0.57
M15	0.88	0.56	0.79	0.57	0.76	0.76	0.87	0.58	0.67	0.53	0.50	0.59
M16												
M17												
M18												
M19												
M21	0.88	0.55	0.78	0.57	0.76	0.76	0.52	0.53	0.54	0.52	0.48	0.53
M22	0.88	0.55	0.78	0.57	0.76	0.76	0.87	0.57	0.66	0.52	0.47	0.59
M23	0.88	0.57	0.79	0.56	0.75	0.75	0.87	0.57	0.66	0.51	0.49	0.58
M24	0.88	0.57	0.79	0.56	0.75	0.75	0.87	0.56	0.65	0.51	0.49	0.58
M25	0.91	0.54	0.76	0.58	0.79	0.79	0.87	0.56	0.68	0.52	0.47	0.59
M26	0.89	0.57	0.79	0.56	0.76	0.76	0.87	0.55	0.67	0.51	0.46	0.58
M27	0.91	0.55	0.72	0.72	0.81	0.81	0.88	0.53	0.70	0.62	0.49	0.64
M28	0.92	0.55	0.76	0.58	0.79	0.79	0.88	0.53	0.68	0.51	0.47	0.60
M29	0.92	0.56	0.79	0.58	0.74	0.74	0.91	0.57	0.65	0.54	0.49	0.56
M30	0.20	0.52		0.50			0.46	0.53		0.49	0.46	
M31												
M32												

References

1. World Health Organization. Ambient Air Pollution: A Global Assessment of Exposure and Burden of Disease. Available online: https://www.who.int/phe/publications/air-pollution-global-assessment/en/ (accessed on 14 January 2020).

2. Fiore, A.M.; Naik, V.; Leibensperger, E.M. Air quality and climate connections. *J. Air Waste Manag. Assoc.* **2015**, *65*, 645–685. [CrossRef] [PubMed]

3. Zhang, Q.; Zheng, Y.; Tong, D.; Shao, M.; Wang, S.; Zhang, Y.; Xu, X.; Wang, J.; He, H.; Liu, W.; et al. Drivers of improved PM2.5 air quality in China from 2013 to 2017. *Proc. Natl. Acad. Sci. USA* **2019**, *116*, 24463–24469. [CrossRef] [PubMed]

4. Ministry of the Environment. Available online: http://www.env.go.jp/air/osen/monitoring.html (accessed on 14 January 2020).

5. Chatani, S.; Yamaji, K.; Sakurai, T.; Itahashi, S.; Shimadera, H.; Kitayama, K.; Hayami, H. Overview of Model Inter-Comparison in Japan's Study for Reference Air Quality Modeling (J-STREAM). *Atmosphere* **2018**, *9*, 19. [CrossRef]

6. Carmichael, G.; Sakurai, T.; Streets, D.; Hozumi, Y.; Ueda, H.; Park, S.; Fung, C.; Han, Z.; Kajino, M.; Engardt, M. MICS-Asia II: The model intercomparison study for Asia Phase II methodology and overview of findings. *Atmos. Environ.* **2008**, *42*, 3468–3490. [CrossRef]

7. Carmichael, G.R.; Calori, G.; Hayami, H.; Uno, I.; Cho, S.Y.; Engardt, M.; Kim, S.-B.; Ichikawa, Y.; Ikeda, Y.; Woo, J.H.; et al. model intercomparison of long-range transport and sulfur deposition in East Asia. *Atmos. Environ.* **2002**, *36*, 175–199. [CrossRef]

8. Chen, L.; Gao, Y.; Zhang, M.; Fu, J.-S.; Zhu, J.; Liao, H.; Li, J.; Huang, K.; Ge, B.; Wang, X.; et al. MICS-Asia III: Multi-model comparison and evaluation of aerosol over East Asia. *Atmos. Chem. Phys. Discuss.* **2019**, *2019*, 1–54. [CrossRef]

9. Li, J.; Nagashima, T.; Kong, L.; Ge, B.; Yamaji, K.; Fu, J.-S.; Wang, X.; Fan, Q.; Itahashi, S.; Lee, H.J.; et al. Model evaluation and inter-comparison of surface-level ozone and relevant species in East Asia in the context of MICS-Asia phase III Part I: Overview. *Atmos. Chem. Phys. Discuss.* **2019**, *2019*, 1–56. [CrossRef]

10. Chatani, S.; Morino, Y.; Shimadera, H.; Hayami, H.; Mori, Y.; Sasaki, K.; Kajino, M.; Yokoi, T.; Morikawa, T.; Ohara, T. Multi-Model Analyses of Dominant Factors Influencing Elemental Carbon in Tokyo Metropolitan Area of Japan. *Aerosol Air Qual. Res.* **2014**, *14*, 396–405. [CrossRef]

11. Shimadera, H.; Hayami, H.; Chatani, S.; Morino, Y.; Mori, Y.; Morikawa, T.; Yamaji, K.; Ohara, T. Sensitivity analyses of factors influencing CMAQ performance for fine particulate nitrate. *J. Air Waste Manag. Assoc.* **2017**, *64*, 374–387. [CrossRef]

12. Shimadera, H.; Hayami, H.; Chatani, S.; Morikawa, T.; Morino, Y.; Mori, Y.; Yamaji, K.; Nakatsuka, S.; Ohara, T. Urban Air Quality Model Inter-Comparison Study (UMICS) for Improvement of PM2.5 Simulation in Greater Tokyo Area of Japan. *Asian J. Atmos. Environ.* **2018**, *12*, 139–152. [CrossRef]

13. Chatani, S.; Yamaji, K.; Itahashi, S.; Saito, M.; Takigawa, M.; Morikawa, T.; Kanda, I.; Miya, Y.; Komatsu, H.; Sakurai, T.; et al. Identifying key factors influencing model performance on ground-level ozone over urban areas in Japan through model inter-comparisons. *Atmos. Environ.* **2020**, *223*. [CrossRef]

14. Itahashi, S.; Yamaji, K.; Chatani, S.; Hayami, H. Refinement of Modeled Aqueous-Phase Sulfate Production via the Fe- and Mn-Catalyzed Oxidation Pathway. *Atmosphere* **2018**, *9*, 132. [CrossRef]

15. Skamarock, W.C.; Klemp, J.B.; Dudhia, J.; Gill, D.O.; Barker, D.M.; Duda, M.G.; Huang, X.Y.; Wang, W.; Powers, J.G. *A Description of the Advanced Research WRF Version 3*; University Corporation for Atmospheric Research: Boulder, CO, USA, 2008.

16. NCEP FNL Operational Model Global Tropospheric Analyses, continuing from July 1999. *Research Data Archive at the National Center for Atmospheric Research*; Computational and Information Systems Laboratory: Boulder, CO, USA, 2000.

17. Gemmill, W.; Katz, B.; Li, X. Daily Real-Time, Global Sea Surface Temperature a High-Resolution Analysis: RTG_SST_I IR, NOAA/NWS/NCEP/EMC/MMAB, Science Application International Corporation, and Joint Center for Satellite Data Assimilation Technical Note Nr. *NOAA/NWS/NCEP/MMAB Off. Note* **2007**, *260*, 1–39.

18. Hong, S.Y.; Dudhia, J.; Chen, S.H. A revised approach to ice microphysical processes for the bulk parameterization of clouds and precipitation. *Mon. Weather Rev.* **2004**, *132*, 103–120. [CrossRef]

19. Mlawer, E.J.; Taubman, S.J.; Brown, P.D.; Iacono, M.J.; Clough, S.A. Radiative transfer for inhomogeneous atmospheres: RRTM, a validated correlated-k model for the longwave. *J. Geophys Res. Atmos* **1997**, *102*, 16663–16682. [CrossRef]

20. Dudhia, J. Numerical Study of Convection Observed during the Winter Monsoon Experiment Using a Mesoscale Two-Dimensional Model. *J. Atmos. Sci.* **1989**, *46*, 3077–3107. [CrossRef]

21. Chen, F.; Dudhia, J. Coupling an advanced land surface-hydrology model with the Penn State-NCAR MM5 modeling system. Part I: Model implementation and sensitivity. *Mon. Weather Rev.* **2001**, *129*, 569–585. [CrossRef]

22. Nakanishi, M.; Niino, H. An improved mellor-yamada level-3 model: Its numerical stability and application to a regional prediction of advection fog. *Bound. Layer Meteorol.* **2006**, *119*, 397–407. [CrossRef]

23. Kain, J.-S. The Kain-Fritsch convective parameterization: An update. *J. Appl. Meteorol.* **2004**, *43*, 170–181. [CrossRef]

24. Byun, D.; Schere, K.L. Review of the Governing Equations, Computational Algorithms, and Other Components of the Models-3 Community Multiscale Air Quality(CMAQ) Modeling System. *Appl. Mech. Rev.* **2006**, *59*, 51–77. [CrossRef]

25. Ramboll Environment and Health. User's Guide Comprehensive Air Quality Model with Extensions. Available online: http://www.camx.com/files/camxusersguide_v6-50.pdf (accessed on 14 January 2020).

26. Grell, G.A.; Peckham, S.E.; Schmitz, R.; McKeen, S.A.; Frost, G.; Skamarock, W.C.; Eder, B. Fully coupled "online" chemistry within the WRF model. *Atmos. Environ.* **2005**, *39*, 6957–6975. [CrossRef]

27. Carter, W.P.L. Documentation of the SAPRC-99 chemical mechanism for VOC reactivity assessment. *Contract* **2000**, *92*, 95–308.

28. Carter, W.P.L. Development of the SAPRC-07 chemical mechanism. *Atmos. Environ.* **2010**, *44*, 5324–5335. [CrossRef]

29. Whitten, G.Z.; Heo, G.; Kimura, Y.; McDonald-Buller, E.; Allen, D.T.; Carter, W.P.L.; Yarwood, G. A new condensed toluene mechanism for Carbon Bond CB05-TU. *Atmos. Environ.* **2010**, *44*, 5346–5355. [CrossRef]

30. Goliff, W.S.; Stockwell, W.R.; Lawson, C.V. The regional atmospheric chemistry mechanism, version 2. *Atmos. Environ.* **2013**, *68*, 174–185. [CrossRef]

31. Binkowski, F.S. Models-3 Community Multiscale Air Quality (CMAQ) model aerosol component 1. Model description. *J. Geophys. Res.* **2003**, 108. [CrossRef]

32. Koo, B.; Knipping, E.; Yarwood, G. 1.5-Dimensional volatility basis set approach for modeling organic aerosol in CAMx and CMAQ. *Atmos. Environ.* **2014**, *95*, 158–164. [CrossRef]

33. Nenes, A.; Pandis, S.N.; Pilinis, C. ISORROPIA: A new thermodynamic equilibrium model for multiphase multicomponent inorganic aerosols. *Aquat. Geochem.* **1998**, *4*, 123–152. [CrossRef]

34. Nenes, A.; Pandis, S.N.; Pilinis, C. Continued development and testing of a new thermodynamic aerosol module for urban and regional air quality models. *Atmos. Environ.* **1999**, *33*, 1553–1560. [CrossRef]

35. Fountoukis, C.; Nenes, A. ISORROPIA II: A computationally efficient thermodynamic equilibrium model for K^+-Ca^{2+}-Mg^{2+}-Nh_4^+-Na^+-SO_4^{2-}-NO_3^--Cl^--H_2O aerosols. *Atmos. Chem. Phys.* **2007**, *7*, 4639–4659. [CrossRef]

36. Carlton, A.G.; Bhave, P.V.; Napelenok, S.L.; Edney, E.D.; Sarwar, G.; Pinder, R.W.; Pouliot, G.A.; Houyoux, M. Model Representation of Secondary Organic Aerosol in CMAQv4.7. *Environ. Sci. Technol.* **2010**, *44*, 8553–8560. [CrossRef]

37. Itahashi, S.; Yamaji, K.; Chatani, S.; Hisatsune, K.; Saito, S.; Hayami, H. Model Performance Differences in Sulfate Aerosol in Winter over Japan Based on Regional Chemical Transport Models of CMAQ and CAMx. *Atmosphere* **2018**, *9*, 488. [CrossRef]

38. Ackermann, I.J.; Hass, H.; Memmesheimer, M.; Ebel, A.; Binkowski, F.S.; Shankar, U. Modal aerosol dynamics model for Europe: Development and first applications. *Atmos. Environ.* **1998**, *32*, 2981–2999. [CrossRef]

39. Schell, B.; Ackermann, I.J.; Hass, H.; Binkowski, F.S.; Ebel, A. Modeling the formation of secondary organic aerosol within a comprehensive air quality model system. *J. Geophys. Res. Atmos* **2001**, *106*, 28275–28293. [CrossRef]

40. Sudo, K.; Takahashi, M.; Kurokawa, J.; Akimoto, H. CHASER: A global chemical model of the troposphere - 1. Model description. *J. Geophys. Res. Atmos.* **2002**, 107. [CrossRef]

41. Huang, M.; Carmichael, G.R.; Pierce, R.B.; Jo, D.S.; Park, R.J.; Flemming, J.; Emmons, L.K.; Bowman, K.W.; Henze, D.K.; Davila, Y.; et al. Impact of intercontinental pollution transport on North American ozone air pollution: An HTAP phase 2 multi-model study. *Atmos. Chem. Phys.* **2017**, *17*, 5721–5750. [CrossRef] [PubMed]

42. Janssens-Maenhout, G.; Crippa, M.; Guizzardi, D.; Dentener, F.; Muntean, M.; Pouliot, G.; Keating, T.; Zhang, Q.; Kurokawa, J.; Wankmuller, R.; et al. HTAP_v2.2: A mosaic of regional and global emission grid maps for 2008 and 2010 to study hemispheric transport of air pollution. *Atmos. Chem. Phys.* **2015**, *15*, 11411–11432. [CrossRef]

43. van der Werf, G.R.; Randerson, J.T.; Giglio, L.; van Leeuwen, T.T.; Chen, Y.; Rogers, B.M.; Mu, M.; van Marle, M.J.E.; Morton, D.C.; Collatz, G.J.; et al. Global fire emissions estimates during 1997–2015. *Earth Syst. Sci. Data Discuss.* **2017**, *2017*, 1–43. [CrossRef]

44. Chatani, S.; Morikawa, T.; Nakatsuka, S.; Matsunaga, S.; Minoura, H. Development of a framework for a high-resolution, three-dimensional regional air quality simulation and its application to predicting future air quality over Japan. *Atmos. Environ.* **2011**, *45*, 1383–1393. [CrossRef]

45. Diehl, T.; Heil, A.; Chin, M.; Pan, X.; Streets, D.; Schultz, M.; Kinne, S. Anthropogenic, biomass burning, and volcanic emissions of black carbon, organic carbon, and SO_2 from 1980 to 2010 for hindcast model experiments. *Atmos. Chem. Phys. Discuss.* **2012**, *2012*, 24895–24954. [CrossRef]

46. Japan Meteorological Agency. Available online: http://www.data.jma.go.jp/svd/vois/data/tokyo/volcano.html (accessed on 14 January 2020).

47. Guenther, A.B.; Jiang, X.; Heald, C.L.; Sakulyanontvittaya, T.; Duhl, T.; Emmons, L.K.; Wang, X. The Model of Emissions of Gases and Aerosols from Nature version 2.1 (MEGAN2.1): An extended and updated framework for modeling biogenic emissions. *Geosci. Model Dev.* **2012**, *5*, 1471–1492. [CrossRef]

48. Tetsuo Fukui, K.K. Tsuyoshi Baba, Akiyoshi Kannari. Updating EAGrid2000-Japan emissions inventory based on the recent emission trends. *J. Jpn. Soc. Atmos. Environ.* **2014**, *49*, 9.

49. Kannari, A.; Tonooka, Y.; Baba, T.; Murano, K. Development of multiple-species 1km×1km resolution hourly basis emissions inventory for Japan. *Atmos. Environ.* **2007**, *41*, 3428–3439. [CrossRef]

50. Zhang, Q.; Streets, D.G.; Carmichael, G.R.; He, K.B.; Huo, H.; Kannari, A.; Klimont, Z.; Park, I.S.; Reddy, S.; Fu, J.-S.; et al. Asian emissions in 2006 for the NASA INTEX-B mission. *Atmos. Chem. Phys.* **2009**, *9*, 5131–5153. [CrossRef]

51. Emmons, L.K.; Walters, S.; Hess, P.G.; Lamarque, J.F.; Pfister, G.G.; Fillmore, D.; Granier, C.; Guenther, A.; Kinnison, D.; Laepple, T.; et al. Description and evaluation of the Model for Ozone and Related chemical Tracers, version 4 (MOZART-4). *Geosci. Model Dev.* **2010**, *3*, 43–67. [CrossRef]
52. Emery, C.; Liu, Z.; Russell, A.G.; Odman, M.T.; Yarwood, G.; Kumar, N. Recommendations on statistics and benchmarks to assess photochemical model performance. *J. Air Waste Manag. Assoc.* **2017**, *67*, 582–598. [CrossRef] [PubMed]
53. Boylan, J.W.; Russell, A.G. PM and light extinction model performance metrics, goals, and criteria for three-dimensional air quality models. *Atmos. Environ.* **2006**, *40*, 4946–4959. [CrossRef]
54. Zhang, Y.; He, J.; Zhu, S.; Gantt, B. Sensitivity of simulated chemical concentrations and aerosol-meteorology interactions to aerosol treatments and biogenic organic emissions in WRF/Chem. *J. Geophys. Res. Atmos.* **2016**, *121*, 6014–6048. [CrossRef]
55. Alexander, B.; Park, R.J.; Jacob, D.J.; Gong, S. Transition metal-catalyzed oxidation of atmospheric sulfur: Global implications for the sulfur budget. *J. Geophys. Res.* **2009**, *114*. [CrossRef]
56. Jacobson, M.Z. Development and application of a new air pollution modeling system.2. Aerosol module structure and design. *Atmos. Environ.* **1997**, *31*, 131–144. [CrossRef]
57. Martin, L.R.; Good, T.W. Catalyzed Oxidation of Sulfur-Dioxide in Solution-The Iron-Manganese Synergism. *Atmos. Environ. Part A Gen. Top.* **1991**, *25*, 2395–2399. [CrossRef]
58. Siefert, R.L.; Johansen, A.M.; Hoffmann, M.R.; Pehkonen, S.O. Measurements of trace metal (Fe, Cu, Mn, Cr) oxidation states in fog and stratus clouds. *J. Air Waste Manag. Assoc.* **1998**, *48*, 128–143. [CrossRef] [PubMed]

Article

Impacts of Biomass Burning Emission Inventories and Atmospheric Reanalyses on Simulated PM_{10} over Indochina

Kyohei Takami [1,2,*]**, Hikari Shimadera** [1]**, Katsushige Uranishi** [1] **and Akira Kondo** [1]

[1] Graduate School of Engineering, Osaka University, 2-1 Yamadaoka, Suita, Osaka 565-0871, Japan; shimadera@see.eng.osaka-u.ac.jp (H.S.); k.uranishi2@gmail.com (K.U.); kondo@see.eng.osaka-u.ac.jp (A.K.)
[2] Environmental Engineering Department, KANSO CO., LTD., 1-3-5 Azuchimachi, Osaka Chuo-ku, Osaka 541-0052, Japan
* Correspondence: takami@ea.see.eng.osaka-u.ac.jp

Received: 19 December 2019; Accepted: 24 January 2020; Published: 4 February 2020

Abstract: Biomass burning (BB) is a major source of atmospheric particles over Indochina during the dry season. Moreover, Indochina has convoluted meteorological scales, and regional meteorological conditions dominate the transport patterns of pollutants. This study focused on the impacts of BB emission inventories and atmospheric reanalyses on simulated PM_{10} over Indochina in 2014 using the Community Multiscale Air Quality (CMAQ) model. Meteorological fields to input to CMAQ were produced by using the Weather Research and Forecasting (WRF) model simulation with the United States National Centers for Environmental Prediction Final (NCEP FNL) Operational Global Analysis or European Centre for Medium Range Weather Forecasts Interim Reanalysis (ERA-interim). The Fire INventory from NCAR (FINN) v1.5 or the Global Fire Emissions Database including small fires (GFED v4.1s) was selected for BB emissions for the air quality simulation. The simulation case with NCEP FNL and FINN v1.5 (FNL + FINN) performed best throughout 2014, including the season when BB activities were intensified. The normalized percentage difference for maximum daily mean PM_{10} concentrations at Chiang Mai for FNL + FINN and the two simulation cases applying GFED v4.1s for BB emissions (−53% to −27%) was much larger than that between the FNL + FINN and ERA + FINN cases (10%). BB emission inventories more strongly impacted PM_{10} simulation than atmospheric reanalyses in highly polluted areas by BB over Indochina in 2014.

Keywords: biomass burning; CMAQ; PM_{10}; atmospheric reanalysis

1. Introduction

Indochina, the peninsular region that includes Cambodia, Laos, Myanmar, Thailand, and Vietnam, has a monsoonal climate with a dry and wet season; the prevailing wind direction and the pattern of precipitation changes seasonally. During the dry season, particulate pollution degrades air quality over Indochina [1]. Outdoor exposure to particulate matter (PM) contributes to ill health, such as cardio- and cerebrovascular disease, respiratory illnesses, lung cancer, and possibly other diseases [2]. A major source of atmospheric particles over Indochina is biomass burning (BB), such as forest fires and agricultural burning. Thailand is surrounded by mountainous areas where BB frequently occurs in the dry season. Chiang Mai, the largest city in northern Thailand, has experienced severe air pollution caused by BB [3–5]. Furthermore, measurements of aerosol properties on Dongsha Island in the northeastern South China Sea revealed that smoke originating from BB in Indochina rises and is trapped within the free troposphere (3–4 km above the earth's surface) in some situations [5–7].

As part of NASA's 2006 Biomass Burning Aerosols in Southeast Asia: Smoke Impact Assessment (BASE-ASIA), regional modeling studies were conducted to assess the impacts of BB in Southeast

Asia on atmospheric composition over Asia in 2006 by using the Community Multiscale Air Quality (CMAQ) model [8]. The CMAQ simulations revealed that during two intense episodes in 2006, BB contributed to ground-level CO, O_3, and $PM_{2.5}$ concentrations in the source region of Southeast Asia by as much as 400 ppbv, 20 ppbv, and 80 µg m^{-3}, respectively [9]. The emissions from BB could also be spread over the southeastern parts of East Asia via strong eastward transport in the free troposphere from 2 to 8 km above the earth's surface [9,10]. It should be noted that the model performance was evaluated using only observed CO concentrations in Thailand, and that discrepancy existed between simulations and measurements due to uncertainty in emission inventories. Amnuaylojaroen et al. [11] reported that BB emissions create a substantial increase for simulated O_3 and CO surface mixing ratios by up to 29% and 16%, respectively, for Southeast Asia in 2008. However, particulate matter was not evaluated in the simulations. Li et al. [12] reported that BB contributed 70%–80% of simulated $PM_{2.5}$ in northern Myanmar, Laos, and Vietnam during March-April 2013.

Several BB emission inventories are available, such as the Fire INventory from NCAR (FINN) [13], the Global Fire Assimilation System (GFAS) [14], and the Global Fire Emissions Database (GFED) [15]. BB emission inventories have substantial uncertainties in their temporal and spatial variabilities because the data to estimate the emissions of BB, such as area, fuel loading, and emission factors, are limited [13,15]. Vongruang et al. [16] reported that PM_{10} was overestimated in air quality simulation with FINN, whereas PM_{10} was underestimated with GFAS in the source region of Northern Thailand in March 2012.

Indochina has convoluted meteorological scales, and regional meteorological conditions dominate the transport patterns of pollutants [1]. Atmospheric reanalyses, initial and boundary conditions for a mesoscale model, have great impacts on the simulated meteorological fields. Several atmospheric reanalyses have been developed by different organizations, for example, the European Centre for Medium Range Weather Forecasts (ECMWF) Interim Reanalysis (ERA-interim) [17], the Japanese 55-year Reanalysis (JRA-55) [18] from the Japan Meteorological Agency (JMA), and the United States National Centers for Environmental Prediction Final (NCEP FNL) [19] Operational Global Analysis. These reanalyses employ various forecast models and data assimilation approaches. ERA-Interim has the highest ability to reproduce climate variables of the East Asian summer monsoon, especially precipitation [20,21].

To the best of our knowledge, there are few studies simulating particulate pollution in Indochina for a period of one year that consider the uncertainties of BB emissions and the complexities of the regional meteorology. In one study, two BB emission inventories were assessed by simulating particulate matter using CMAQ during a specific pollution event [16]. This study focused on the impacts of BB emission inventories and atmospheric reanalyses on the simulation of PM_{10} over Indochina in 2014. From the end of February to early April in 2014, agricultural burning and forest fires in northern Thailand caused severe particulate pollution over the region [5].

2. Materials and Methods

2.1. Simulation Design

Air quality simulations for the year 2014 were conducted by CMAQ v5.0.2 with an initial spin up period of December 2013. The modeling domains covered regions from East Asia (D1) to Indochina (D2) (Figure 1). The horizontal resolutions were 72 km and 24 km, and the number of grid cells were 92 × 92 and 98 × 98 for D1 and D2, respectively. The domains consisted of 30 sigma-pressure coordinate vertical layers ranging from the surface to 100 hPa. Figure 2 shows the procedure of air quality simulation in this study.

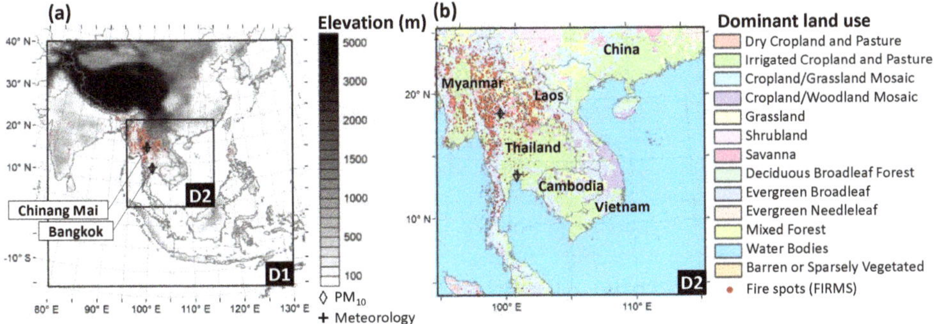

Figure 1. Locations of observation sites for PM$_{10}$ and meteorology, and fire spots (MCD14DL) provided by Fire Information for Resource Management System (FIRMS) on 21 March 2014 in the modeling domains covering (**a**) East Asia (D1) and (**b**) Indochina (D2). Elevation and dominant United States Geological Survey (USGS) 24-category land use are provided in (**a**) and (**b**), respectively.

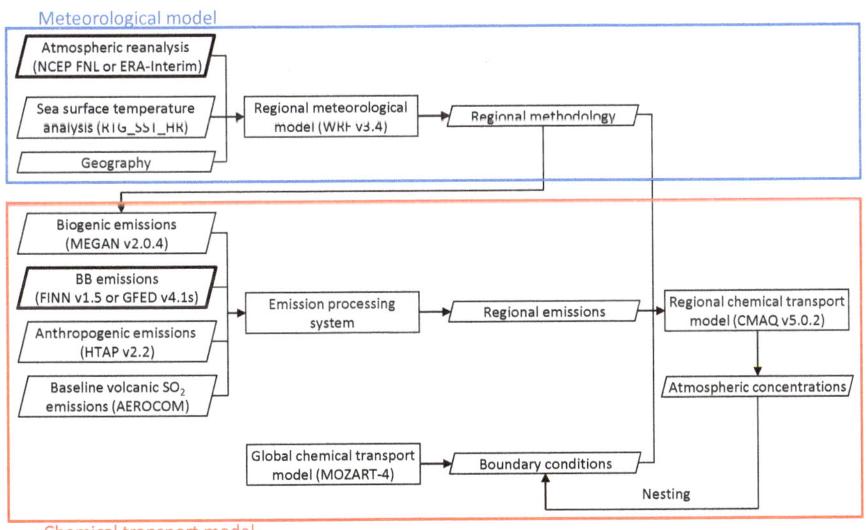

Figure 2. Procedure of air quality simulation by meteorological model and chemical transport model.

2.2. Meteorological Model

Meteorological fields to input to CMAQ were produced by using the Weather Research and Forecasting (WRF) model [22] v3.4. The WRF simulations were conducted using the high-resolution, real-time, global sea surface temperature analysis (RTG_SST_HR) of NCEP, and two atmospheric reanalyses: NCEP FNL or ERA-Interim. The physics options were as follows: Rapid Radiative Transfer Model for General Circulation Models Shortwave and Longwave Schemes [23], planetary boundary layer physics of Asymmetric Convection Model 2 Scheme [24], cumulus parameterization of the Kain–Fritsch Scheme [25], micro physics of Morrison 2-moment Scheme [26], and Pleim–Xiu Land Surface Model [27]. Grid nudging was applied to horizontal wind components at all the vertical layers with a coefficient of 3.0×10^{-4} s^{-1} in D1 and 1.0×10^{-4} s^{-1} in D2 for the entire period.

2.3. Chemical Transport Model

CMAQ was configured with the Carbon Bond chemical mechanism (CB05) [28] for gas-phase chemistry and the sixth generation CMAQ aerosol module for the aerosol process. The initial and lateral boundary conditions for D1 were obtained from the Model for Ozone and Related Chemical Tracers v4 (MOZART-4) [29]. Several datasets were used to produce emission data. In particular, the Fire INventory from NCAR (FINN) [11] v1.5 or the Global Fire Emissions Database including small fires (GFED v4.1s) [30] was selected for BB emissions. FINN v1.5 provided 1 km-resolution BB emissions estimated from the active fire observations from the Moderate Resolution Imaging Spectroradiometer (MODIS) Terra and Aqua satellite. GFED v4.1s provided 0.25°-resolution BB emissions estimated from the 500 m Collection 5.1 MODIS direct broadcast burned area product plus additional burned area from small fires based on a revised version of the Randerson et al. [31] small-fire estimation approach. Anthropogenic emissions were obtained from the Task Force Hemispheric Transport of Air Pollution (HTAP) v2.2 inventory [32]. The other natural emissions were derived from the Model of Emissions of Gases and Aerosols from Nature (MEGAN) [33] v2.04 for biogenic emissions and the Aerosol Comparisons between Observations and Models (AEROCOM) data [34] for baseline volcanic SO_2 emissions. There were differences between the periods of simulations and the reference years in the anthropogenic emission data available for the simulations, and the uncertainties associated with the differences may have affected model performance.

In order to evaluate the impacts of BB emission inventories and atmospheric reanalyses on simulated PM_{10} concentrations, four cases of simulations (FNL + FINN; FNL + GFED; ERA + FINN; and ERA + GFED) were executed. The first parts of cases' names indicate the applied reanalysis: FNL and ERA stand for NCEP FNL and ERA-Interim, respectively. The second parts indicate the selected BB emission inventories. Furthermore, simulations without BB emission inventories (FNL + exBB; ERA + exBB) were conducted. The difference between cases with BB emissions and those without BB emissions was calculated to estimate BB contributions to PM_{10} concentrations in the four simulation cases with BB emissions.

Figure 3 shows spatial distributions of PM_{10} emissions from FINN v1.5 and GFED v4.1s in March and April 2014. During these two months, PM_{10} emissions in both BB emission inventories were concentrated in Indochina, especially the mountainous areas in Laos, Myanmar, and Thailand. FINN v1.5 estimated larger emissions in a wider area than GFED v4.1s. Figure 4 shows the variation of the amount of PM_{10} emissions over D2 and the relative contributions from BB. PM_{10} emissions from BB produced by FINN v1.5 and GFED v4.1s accounted for 244 mg s^{-1} km^{-2} and 91 mg s^{-1} km^{-2}, respectively, which contributed 93% and 83%, respectively, of the total amount of PM_{10} emissions in March when burning activities were intensified.

The WRF performance was evaluated through comparison with ground-level air temperature, relative humidity, and wind speed data in Bangkok and Chiang Mai distributed by the University of Wyoming [35]. The CMAQ performance was evaluated through comparison with ground-level PM_{10} concentration data in Bangkok and Chiang Mai distributed by the Acid Deposition Monitoring Network in East Asia (EANET) [36].

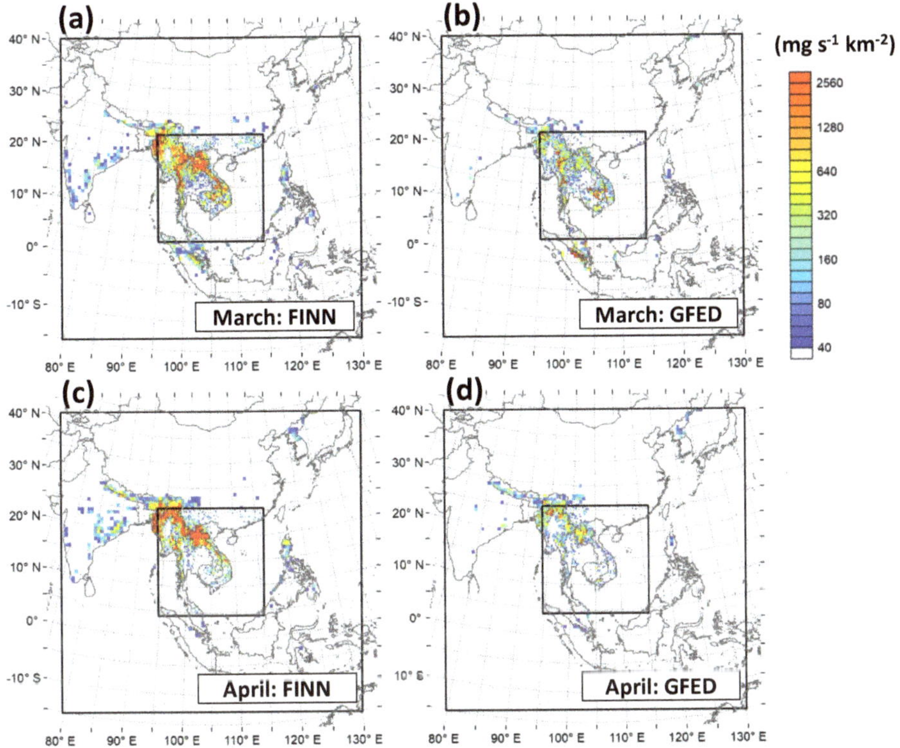

Figure 3. Spatial distributions of mean PM_{10} emissions from biomass burning: (**a**) FINN v1.5 and (**b**) GFED v4.1s in March, and (**c**) FINN v1.5 and (**d**) GFED v4.1s in April.

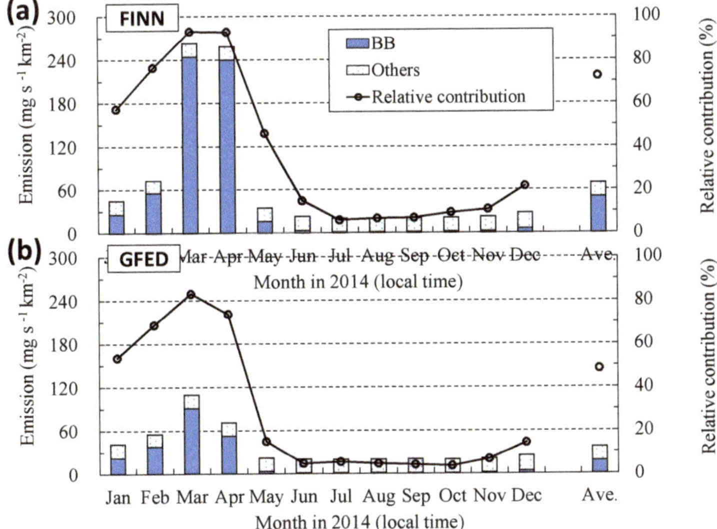

Figure 4. Variation of mean PM_{10} emissions over D2 and the relative contributions from biomass burning (BB): (**a**) FINN v1.5 and (**b**) GFED v4.1s.

3. Results

3.1. Daily Meteorological Factors at Observation Sites

The WRF performance was evaluated by using the normalized mean bias (NMB) and the index of agreement (IA) for ground-level air temperature, relative humidity, and wind speed for Bangkok and Chiang Mai (Table 1). More statistics were shown in supplementary data (Table S1). IA is a statistical measure to present the model performance and varies between 0 and 1, where 1 means model performance is perfect [37]. IA values for air temperature, relative humidity, and wind speed were relatively high in both WRF simulations: the models applying both NCEP FNL and ERA-interim simulated spatial and temporal variations of these parameters in the same level.

Table 1. Model performance for daily mean meteorological factors in the two simulation cases for the periods of one year and two months (from March to April) in 2014.

	Site	FNL		ERA		Observation	
		Year	Mar–Apr	Year	Mar–Apr	Year	Mar–Apr
Temperature							
Mean	Bangkok	27	30	27	30	29	30
(°C)	Chiang Mai	26	31	26	31	26	28
NMB	Bangkok	−5	−1	−5	−1		
(%)	Chiang Mai	−2	8	−2	9		
IA	Bangkok	0.76	0.75	0.79	0.74		
	Chiang Mai	0.86	0.62	0.85	0.60		
Relative Humidity							
Mean	Bangkok	78	70	77	70	70	70
(%)	Chiang Mai	68	37	66	34	64	47
NMB	Bangkok	11	0	9	0		
(%)	Chiang Mai	6	−21	2	−28		
IA	Bangkok	0.65	0.57	0.67	0.56		
	Chiang Mai	0.78	0.60	0.79	0.54		
Wind Speed							
Mean	Bangkok	2.4	2.9	2.5	2.9	3.2	3.7
(m s^{-1})	Chiang Mai	1.7	2.1	1.8	2.1	1.9	2.1
NMB	Bangkok	−27	−21	−22	−19		
(%)	Chiang Mai	−11	0	−8	0		
IA	Bangkok	0.61	0.66	0.62	0.67		
	Chiang Mai	0.58	0.64	0.59	0.70		

Note: Mean—mean value, NMB—normalized mean bias, IA—index of agreement.

3.2. Daily Concentrations of PM$_{10}$ at Observation Sites

The performance of daily mean PM$_{10}$ concentration simulations in the four simulation cases with BB emissions was evaluated by using NMB and IA for the periods of both the entire year and the two month period from March to April 2014 (Table 2). More statistics were shown in supplementary data (Table S2). The highest PM$_{10}$ concentrations in 2014 were observed on January 3 at Bangkok (123.9 µg m^{-3}) and on March 21 at Chiang Mai (242.9 µg m^{-3}), respectively. At Chiang Mai, maximum daily mean PM$_{10}$ concentrations were simulated in all of the four simulation cases on the same day: March 21. The IA value for FNL + FINN was the highest for both the one-year and the two-month periods, indicating that FNL + FINN performed best throughout 2014, including the season when BB activities were intensified.

Table 2. Model performance for daily mean PM_{10} concentrations in the four simulation cases for the periods of one full year (2014) and two months (from March to April 2014).

	Site	FNL + FINN		FNL + GFED		ERA + FINN		ERA + GFED		Observation	
		Year	Mar–Apr	Year	Mar–Apr	Year	Mar–Apr	Year	Mar–Apr	Year	Mar–Apr
Mean	Bangkok	32.2	32.2	30.7	27.3	31.0	35.1	29.3	29.3	37.2	38.8
(μg m^{-3})	Chiang Mai	37.2	94.1	30.1	57.6	40.1	107.3	32.3	66.7	46.6	87.9
Maximum	Bangkok	107.2	93.9	101.7	65.7	104.9	104.9	91.2	73.9	123.9	77.1
(μg m^{-3})	Chiang Mai	309.5	309.5	145.8	145.8	339.5	339.5	225.4	225.4	242.9	242.9
NMB	Bangkok	−13	−17	−17	−30	−17	−9	−21	−24		
(%)	Chiang Mai	−20	7	−35	−34	−14	22	−31	−24		
IA	Bangkok	0.88	0.84	0.85	0.78	0.85	0.83	0.82	0.78		
	Chiang Mai	0.84	0.77	0.75	0.67	0.83	0.73	0.80	0.74		

Note: Mean—mean value, Maximum—maximum daily mean value, NMB—normalized mean bias, IA—index of agreement.

Figure 5 shows simulated and observed daily concentrations at Bangkok and Chiang Mai. All of the four simulation cases simulated the day-to-day variation patterns well at both sites for the year 2014. Cases applying FINN v1.5 for BB emissions (FNL + FINN; ERA + FINN) showed higher trends of simulated PM_{10} concentrations at both observation sites compared to those applying GFED v4.1s (FNL + GFED; ERA + GFED), regardless of the atmospheric reanalysis. In particular, the former two cases clearly presented higher trends of PM_{10} concentrations than the latter cases at Chiang Mai from the end of February to early April, when high PM_{10} concentrations were observed. Table 3 shows the normalized percentage difference for FNL + FINN and the other three cases for annual mean and maximum daily mean PM_{10} concentrations in 2014 at the two observation sites. The percentage difference at Bangkok (−9% to −4% for annual mean; −15% to −2% for maximum) was less than that at Chiang Mai (−19%–8% for annual mean; −53%–10% for maximum). At both of the two sites, the percentage difference between the FNL + FINN and ERA + FINN cases was the smallest for annual mean and maximum daily mean PM_{10} concentrations in 2014. In particular, the difference for maximum daily mean PM_{10} concentrations at Chiang Mai for FNL + FINN and the two cases applying GFED v4.1s for BB emissions (−53% to −27%) was much larger than that between the FNL + FINN and ERA + FINN cases (10%). Consequently, BB emission inventories more strongly impacted the PM_{10} simulation than atmospheric reanalyses at the two sites in Indochina.

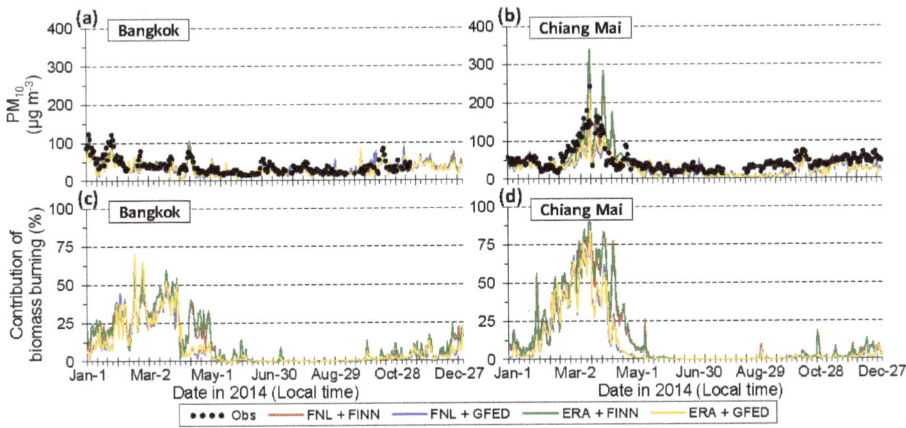

Figure 5. Time series of simulated and observed daily mean concentrations of PM_{10} at (**a**) Bangkok and (**b**) Chiang Mai and estimated daily mean relative contributions of BB at (**c**) Bangkok and (**d**) Chiang Mai.

Table 3. Normalized percentage difference for FNL + FINN and the other three cases for daily mean PM$_{10}$ concentrations for the year 2014.

	Site	FNL + GFED	ERA + FINN	ERA + GFED
FNL + FINN				
Mean (%)	Bangkok	−5	−4	−9
	Chiang Mai	−19	8	−13
Maximum (%)	Bangkok	−5	−2	−15
	Chiang Mai	−53	10	−27

Note: Mean—annual mean value, Maximum—maximum daily mean value.

BB contributions were estimated by the difference between cases with and without BB emissions (Figure 5). The BB contributions to PM$_{10}$ concentrations of the four simulation cases were 2%–12% at Bangkok and 75%–89% at Chiang Mai for each day that the maximum concentrations were observed. Li et al. [12] reported that BB contributions to simulated PM$_{2.5}$ was 70%–80% in the BB source regions during March-April 2013 and the contributions were close to 75%–89% at Chiang Mai on 21 March 2014.

Figure 6 shows simulated and observed wind speed, planetary boundary layer (PBL) height, and daily concentrations at Bangkok and Chiang Mai from March to April. At Bangkok, these parameters were on the similar trends among the four simulation cases. At Chiang Mai, the maximum PM$_{10}$ concentration was observed in weak-wind condition. On the other hand, there was little relationship between PBL height and PM$_{10}$ concentrations. The simulated spatial distributions on January 3 and March 21 are analyzed in detail in the next subsection.

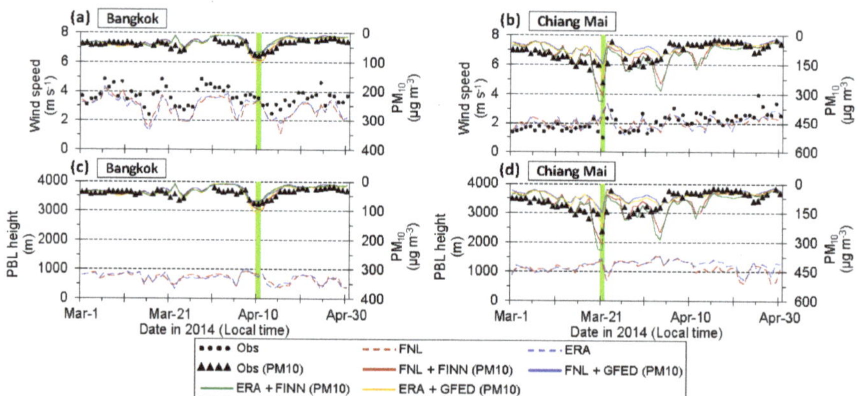

Figure 6. Time series of simulated and observed daily mean wind speed at (**a**) Bangkok and (**b**) Chiang Mai, and simulated daily mean planetary boundary layer (PBL) height at (**c**) Bangkok and (**d**) Chiang Mai. The simulated and observed daily mean concentrations of PM$_{10}$ were also shown. The day when maximum PM$_{10}$ concentration was observed from March to April 2014 at each site was highlighted.

3.3. Spatial Distributions for Daily Concentrations of PM$_{10}$

Figure 7 shows the spatial distributions of daily PM$_{10}$ concentrations and wind fields at ground level on January 3, when maximum PM$_{10}$ concentration was observed in Bangkok. The large BB contribution area was spread out over Southern China and Cambodia. On January 3, ground-level winds were weak over Indochina. The area where simulated PM$_{10}$ concentrations were around 80 µg m^{-3} was spread across Thailand, and the BB contributions were relatively small in the four cases; discrepancy among the four cases was small across Thailand.

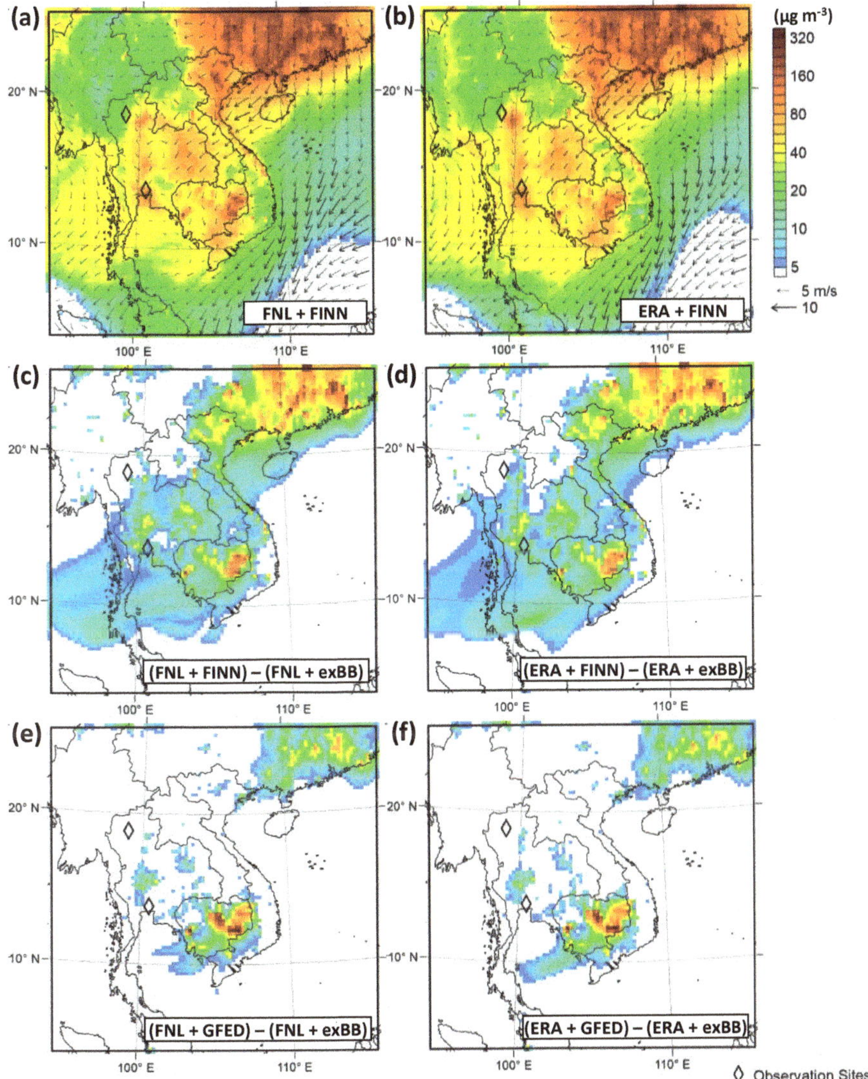

Figure 7. Spatial distributions of daily mean of PM_{10} concentrations and wind fields at ground level in (**a**) FNL + FINN and (**b**) ERA + FINN cases, and BB contributions to PM_{10} concentrations in (**c**) FNL + FINN, (**d**) ERA + FINN, (**e**) FNL + GFED, and (**f**) ERA + GFED cases on 3 January 2014.

In March 2014, the largest number of hotspots were detected in Thailand and Myanmar from March 18 to 21 [5]. A large number of fire spots occurred in the mountainous areas of Indochina on March 21 (Figure 1). Figure 8 shows spatial distributions of daily PM_{10} concentrations and wind fields at ground level on March 21. Ground-level winds blew easterly in the eastern part of Indochina and converged at the Myanmar and Thailand border in all cases. Concentrations of 160 $\mu g\ m^{-3}$ of PM_{10} were spread over the area along the Myanmar and Thailand border in the four cases with BB emissions; the two cases that FINN v1.5 selected for BB emissions were predicted to have higher PM_{10} concentrations in and around Northern Laos. In highly polluted areas caused by BB over Indochina, the discrepancy of simulated PM_{10} concentrations resulting from different BB emission inventories was larger than that resulting from different atmospheric reanalyses.

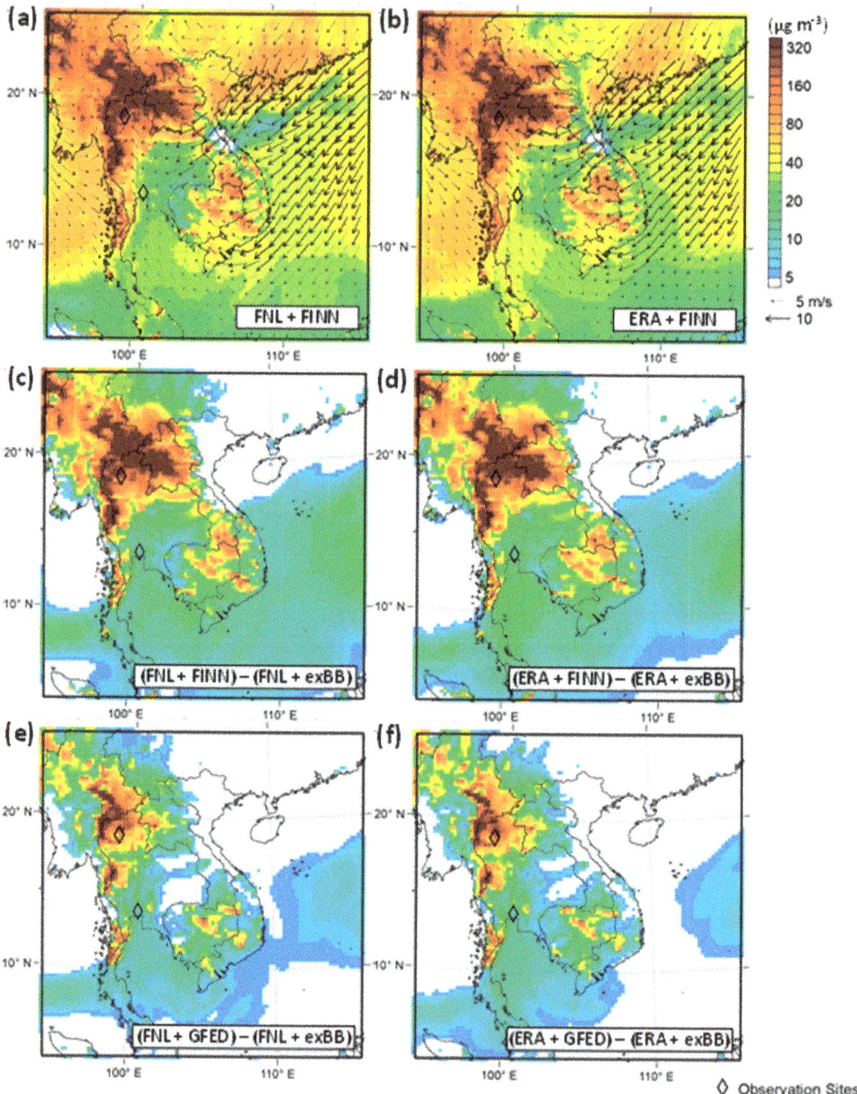

Figure 8. Spatial distributions of daily mean of PM_{10} concentrations and wind fields at ground level in (**a**) FNL + FINN and (**b**) ERA + FINN cases, and BB contributions to PM_{10} concentrations in (**c**) FNL + FINN, (**d**) ERA + FINN, (**e**) FNL + GFED, and (**f**) ERA + GFED cases on 21 March 2014.

4. Conclusions

This study focused on the impacts of BB emission inventories and atmospheric reanalyses on the simulation of PM_{10} over Indochina for the year 2014. Meteorological fields to input to CMAQ were produced by using the WRF model simulation with NCEP FNL or ERA-interim. FINN v1.5 or GFED v4.1s were selected for BB emissions. In order to evaluate the impacts, four cases of simulations (FNL + FINN; FNL + GFED; ERA + FINN; ERA + GFED) were executed.

The performance of CMAQ simulations for PM_{10} concentrations at ground-level was evaluated at Bangkok and Chiang Mai in Thailand. The FNL + FINN case performed best for both the full year and the two-month period when BB activities were intensified in 2014. To compare the impacts of input data (BB emission inventories and atmospheric reanalyses) on simulated PM_{10}, we introduced the normalized percentage difference for FNL + FINN and the other three cases (FNL + GFED; ERA + FINN; ERA + GFED) for annual mean and maximum daily mean PM_{10} concentrations in 2014 at the two observation sites. At both of the two sites, the normalized percentage difference for annual mean and maximum daily mean PM_{10} concentrations in 2014 for FNL + FINN and the two cases applying GFED v4.1s for BB emissions (FNL + GFED; ERA + GFED) was larger than that between FNL + FINN and ERA + FINN cases. In particular, the difference for maximum daily mean PM_{10} concentrations at Chiang Mai for FNL + FINN and the two cases applying GFED v4.1s (-53% to -27%) was much larger than that between the FNL + FINN and ERA + FINN cases (10%).

BB contributions were estimated by the difference between the cases with BB emissions and those without BB emissions. BB contributions to PM_{10} concentrations in the four simulation cases at Chiang Mai (75%–89%) were larger than those at Bangkok (2%–12%) on the day when the highest PM_{10} concentrations were observed in 2014 at each observation site. BB highly contributed on the particulate pollution over Southern china and Cambodia on January 3, and over the area along the Myanmar and Thailand border on March 21 in weak-wind condition. In polluted areas caused by BB, the cases applying FINN v1.5 for BB emission inventories showed higher trends of PM_{10} concentrations. Selected BB emission inventories more strongly impacted the PM_{10} simulation than selected atmospheric reanalyses in highly polluted areas by BB over Indochina in 2014.

This approach is applicable to other years and atmospheric pollutants to evaluate the impacts of selected input datasets on air quality simulations in polluted areas caused by BB. For example, GFAS can be used for one of the other BB emission inventories. However, emission factors of global BB emission inventories, such as FINN, GFAS, and GFED, were determined based on generic vegetation types of each fire pixel. BB-emission data focusing on the unique characteristics of BB in the specific region would reduce the uncertainties of BB emissions. We hope that this study will help to improve understanding about the mechanisms that degrade air quality over Indochina.

Supplementary Materials: The following are available online at http://www.mdpi.com/2073-4433/11/2/160/s1, Table S1. Model performance for daily mean meteorological factors in the two simulation cases for the periods of one year and two months (from March to April) in 2014. Table S2. Model performance for daily mean PM10 concentrations in the four simulation cases for the periods of one full year (2014) and two months (from March to April 2014).

Author Contributions: Conceptualization, K.T., H.S., and A.K.; methodology, H.S. and K.U.; software, H.S. and K.U.; validation, K.T. and H.S.; formal analysis, K.T. and H.S.; data curation, H.S.; writing—original draft preparation, K.T.; writing—review and editing, H.S., K.U., and A.K.; visualization, K.T.; supervision, A.K.; project administration, H.S. All authors have read and agreed to the published version of the manuscript.

Funding: This research received no external funding.

Conflicts of Interest: The authors declare no conflict of interest.

References

1. Reid, J.S.; Hyer, E.J.; Johnson, R.S.; Holben, B.N.; Yokelson, R.J.; Zhang, J.; Campbell, J.R.; Christopher, S.A.; Girolamo, L.D.; Giglio, L.; et al. Observing and understanding the Southeast Asian aerosol system by remote sensing: An initial review and analysis for the Seven Southeast Asian Studies (7SEAS) program. *Atmos. Res.* **2013**, *122*, 403–468. [CrossRef]

2. IPCC. Climate Change 2014: Impacts, Adaptation, and Vulnerability. Part A: Global and Sectoral Aspects. Working Group II Contribution to the Fifth Assessment Report of the Intergovernmental Panel on Climate Change. Available online: https://www.ipcc.ch/report/ar5/wg2/ (accessed on 19 December 2019).

3.   Pengchai, P.; Chantara, S.; Sopajaree, K.; Wangkarn, S.; Tencharoenkul, U.; Rayanakorn, M. Seasonal variation, risk assessment and source estimation of PM 10 and PM10-bound PAHs in the ambient air of Chiang Mai and Lamphun, Thailand. *Environ. Monit. Assess.* **2009**, *154*, 197–218. [CrossRef] [PubMed]

4.   Tsai, Y.I.; Soparajee, K.; Chotruksa, A.; Wu, H.C. Source indicators of biomass burning associated with inorganic salts and carboxylates in dry season ambient aerosol in Chiang Mai basin, Thailand. *Atmos. Environ.* **2013**, *78*, 93–104. [CrossRef]

5.   Duc, H.N.; Bang, H.Q.; Quang, N.X. Modelling and prediction of air pollutant transport during the 2014 biomass burning and forest fires in peninsular Southeast Asia. *Environ. Monit. Assess.* **2016**, *188*, 106. [CrossRef] [PubMed]

6.   Atwood, S.A.; Reid, J.S.; Kreidenweis, S.M.; Cliff, S.; Zhao, Y.; Lin, N.-H.; Tsay, S.-C.; Chu, Y.-C.; Westphal, D.L. Size resolved measurements of springtime aerosol particles over the northern South China Sea. *Atmos. Environ.* **2013**, *78*, 134–143. [CrossRef]

7.   Wang, S.H.; Tsay, S.-C.; Lin, N.-H.; Chang, S.-C.; Li, C.; Welton, E.J.; Holben, B.N.; Hsu, N.C.; Lau, W.K.-M.; Lolli, S.; et al. Origin, transport, and vertical distribution of atmospheric pollutants over the northern South China Sea during 7-SEAS/Dongsha experiment. *Atmos. Environ.* **2013**, *78*, 124–133. [CrossRef]

8.   Byun, D.W.; Schere, K.L. Review of the governing equations, computational algorithms, and other components of the Models-3 community multiscale air quality (CMAQ) modeling system. *Appl. Mech. Rev.* **2006**, *59*, 51–77. [CrossRef]

9.   Fu, J.S.; Hsu, N.C.; Gao, Y.; Huang, K.; Li, C.; Lin, N.-H.; Tsay, S.-C. Evaluating the influences of biomass burning during 2006 BASE-ASIA: A regional chemical transport modeling. *Atmos. Chem. Phys.* **2012**, *12*, 3837–3855. [CrossRef]

10.   Huang, K.; Fu, J.S.; Hsu, N.C.; Gao, Y.; Dong, X.; Tsay, S.-C.; Lam, Y.F. Impact assessment of biomass burning on air quality in Southeast and East Asia during BASE-ASIA. *Atmos. Environ.* **2013**, *78*, 291–302. [CrossRef]

11.   Amnuaylojaroen, T.; Barth, M.C.; Emmons, L.K.G.; Carmichael, R.; Kreasuwun, J.; Prasitwattanaseree, S.; Chantara, S. Effect of different emission inventories on modeled ozone and carbon monoxide in Southeast Asia. *Atmos. Environ.* **2014**, *14*, 12983–13012.

12.   Li, J.; Zhang, Y.; Wang, Z.; Sun, Y.; Fu, P.; Yang, Y.; Huang, H.; Li, J.; Zhang, Q.; Lin, C.; et al. Regional impact of biomass burning in Southeast Asia on atmospheric aerosols during the 2013 seven South-East Asian studies project. *Aerosol Air Qual. Res.* **2017**, *17*, 2924–2941. [CrossRef]

13.   Wiedinmyer, C.; Akagi, S.K.; Yokelson, R.J.; Emmons, L.K.; Al-Saadi, J.A.; Orlando, J.J.; Soja, A.J. The Fire INventory from NCAR (FINN): A high resolution global model to estimate the emissions from open burning. *Geosci. Model Dev.* **2011**, *4*, 625–641. [CrossRef]

14.   Kaiser, J.W.; Heil, A.; Andreae, M.O.; Benedetti, A.; Chubarova, N.; Jones, L.; Morcrette, J.-J.; Razinger, M.; Schultz, M.G.; Suttie, M.; et al. Biomass burning emissions estimated with a global fire assimilation system based on observed fire radiative power. *Biogeosciences* **2012**, *9*, 527–554. [CrossRef]

15.   Giglio, L.; Randerson, J.T.; van der Werf, G.R. Analysis of daily, monthly, and annual burned area using the fourth-generation global fire emissions database (GFED4). *J. Geophys. Res. Biogeosci.* **2013**, *118*, 317–328. [CrossRef]

16.   Vongruang, P.; Wongwises, P.; Pimonsree, S. Assessment of fire emission inventories for simulating particulate matter in Upper Southeast Asia using WRF-CMAQ. *Atmos. Pollut. Res.* **2017**, *8*, 921–929. [CrossRef]

17.   Dee, D.P.; Uppala, S.M.; Simmons, A.J.; Berrisford, P.; Poli, P.; Kobayashi, S.; Andrae, U.; Balmaseda, M.A.; Balsamo, G.; Bauer, P.; et al. The ERA-Interim reanalysis: Configuration and performance of the data assimilation system. *Q. J. R. Meteorolog. Soc.* **2011**, *137*, 553–597. [CrossRef]

18.   Kobayashi, S.; Ota, Y.; Harada, Y.; Ebita, A.; Moriya, M.; Onoda, H.; Onogi, K.; Kamahori, H.; Kobayashi, C.; Endo, H.; et al. The JRA-55 reanalysis: General specifications and basic characteristics. *J. Meteorol. Soc. Jpn. Ser. II* **2015**, *93*, 5–48. [CrossRef]

19.   Kalnay, E.; Kanamitsu, M.; Kistler, R.; Collins, W.; Deaven, D.; Gandin, L.; Iredell, M.; Saha, S.; White, G.; Woollen, J.; et al. The NCEP/NCAR 40-year reanalysis project. *Bull. Am. Meteorol. Soc.* **1996**, *77*, 437–471. [CrossRef]

20.   Huang, D.-Q.; Zhu, J.; Zhang, Y.-C.; Huang, Y.; Kuang, X.-Y. Assessment of summer monsoon precipitation derived from five reanalysis datasets over East Asia. *Q. J. R. Meteorol. Soc.* **2016**, *142*, 108–119. [CrossRef]

21.   Huang, B.; Cubasch, U.; Li, Y. East Asian summer monsoon representation in re-analysis datasets. *Atmosphere* **2018**, *9*, 235. [CrossRef]

22. Skamarock, W.C.; Klemp, J.B. A time-split nonhydrostatic atmospheric model for weather research and forecasting applications. *J. Comput. Phys.* **2008**, *227*, 3465–3485. [CrossRef]

23. Iacono, M.J.; Delamere, J.S.; Mlawer, E.J.; Shephard, M.W.; Clough, S.A.; Collins, W.D. Radiative forcing by long-lived greenhouse gases: Calculations with the AER radiative transfer models. *J. Geophys. Res.* **2008**, *113*, D13103. [CrossRef]

24. Pleim, J.E. A combined local and nonlocal closure model for the atmospheric boundary layer. Part I: Model description and testing. *J. Appl. Meteor. Climatol.* **2007**, *46*, 1383–1395. [CrossRef]

25. Kain, J.S. The Kain-Fritsch convective parameterization: An update. *J. Appl. Meteorol.* **2004**, *43*, 170–181. [CrossRef]

26. Morrison, H.; Thompson, G.; Tatarskii, V. Impact of cloud microphysics on the development of trailing stratiform precipitation in a simulated squall line: Comparison of one– and two–moment schemes. *Mon. Wea. Rev.* **2009**, *137*, 991–1007. [CrossRef]

27. Xiu, A.; Pleim, J.E. Development of a land surface model. Part I: Application in a mesoscale meteorological model. *J. Appl. Meteor.* **2001**, *40*, 192–209. [CrossRef]

28. Yarwood, G.; Rao, S.; Yocke, M.; Whitten, G.Z. Updates to the carbon bond chemical mechanism: CB05. *Final Rep. US EPA* **2005**, *8*, 13.

29. Emmons, L.K.; Walters, S.; Hess, P.G.; Lamarque, J.-F.; Pfister, G.G.; Fillmore, D.; Kloster, S. Description and evaluation of the Model for Ozone and Related chemical Tracers, version 4 (MOZART-4). *Geosci. Model Dev.* **2010**, *3*, 43–67. [CrossRef]

30. van der Werf, G.R.; Randerson, J.T.; Giglio, L.; van Leeuwen, T.T.; Chen, Y.; Rogers, B.M.; Mu, M.; van Marle, M.J.E.; Morton, D.C.; Collatz, G.J.; et al. Global fire emissions estimates during 1997–2016. *Earth Syst. Sci. Data* **2017**, *9*, 697–720. [CrossRef]

31. Randerson, J.T.; Chen, Y.; van der Werf, G.R.; Rogers, B.M.; Morton, D.C. Global burned area and biomass burning emissions from small fires. *J. Geophys. Res.* **2012**, *117*, G04012. [CrossRef]

32. Janssens-Maenhout, G.; Crippa, M.; Guizzardi, D.; Dentener, F.; Muntean, M.; Pouliot, G.; Li, M. HTAP-v2.2: A mosaic of regional and global emission grid maps for 2008 and 2010 to study hemispheric transport of air pollution. *Atmos. Chem. Phys.* **2015**, *15*, 11411–11432. [CrossRef]

33. Guenther, A.; Karl, T.; Harley, P.; Wiedinmyer, C.; Palmer, P.I.; Geron, C. Estimates of global terrestrial isoprene emissions using MEGAN (model of emissions of gases and aerosols from nature). *Atmos. Chem. Phys.* **2006**, *6*, 3181–3210. [CrossRef]

34. Diehl, T.; Heil, A.; Chin, M.; Pan, X.; Streets, D.; Schultz, M.; Kinne, S. Anthropogenic, biomass burning, and volcanic emissions of black carbon, organic carbon, and SO_2 from 1980 to 2010 for hindcast model experiments. *Atmos. Chem. Phys. Discuss.* **2012**, *12*, 24895–24954. [CrossRef]

35. University of Wyoming, Wyoming Weather Web. Available online: http://weather.uwyo.edu/ (accessed on 19 December 2019).

36. Network Center for EANET, EANET Data on the Acid Deposition in the East Asian Region. Available online: https://monitoring.eanet.asia/document/public/index (accessed on 14 June 2019).

37. Willmott, C.J. On the validation of models. *Phys. Geogr.* **1981**, *2*, 184–194. [CrossRef]

Article

RePLaT-Chaos: A Simple Educational Application to Discover the Chaotic Nature of Atmospheric Advection

Tímea Haszpra

Department of Theoretical Physics, and MTA–ELTE Theoretical Physics Research Group,
H-1117 Budapest, Hungary; hatimi@caesar.elte.hu

Received: 10 November 2019; Accepted: 21 December 2019; Published: 27 December 2019

Abstract: Large-scale atmospheric pollutant spreading via volcano eruptions and industrial accidents may have serious effects on our life. However, many students and non-experts are generally not aware of the fact that pollutant clouds do not disperse in the atmosphere like dye blobs on clothes. Rather, an initially compact pollutant cloud soon becomes strongly stretched with filamentary and folded structure. This is the result of the chaotic behaviour of advection of pollutants in 3-D flows, i.e., the advection dynamics of pollutants shows the typical characteristics such as sensitivity to the initial conditions, irregular motion, and complicated but well-organized (fractal) structures. This study presents possible applications of a software called RePLaT-Chaos by means of which the characteristics of the long-range atmospheric spreading of volcanic ash clouds and other pollutants can be investigated in an easy and interactive way. This application is also a suitable tool for studying the chaotic features of the advection and determines two quantities which describe the chaoticity of the advection processes: the stretching rate quantifies the strength of the exponential stretching of pollutant clouds; and the escape rate characterizes the rate of the rapidity by which the settling particles of a pollutant cloud leave the atmosphere.

Keywords: RePLaT-Chaos; large-scale atmospheric advection; chaotic advection; stretching rate; escape rate; education

---

1. Introduction

Air pollution is an important environmental issue, especially, in cases when pollutants travel thousands of kilometers, affecting air quality far away from their initial source. In the last decade several events drew even non-specialists' attention to the potential continental and global impacts of pollutant emissions from natural sources or anthropogenic industrial accidents. For example, in April and May 2010 the eruptions of the Icelandic Eyjafjallajökull volcano resulted in airspace closures across Europe. As a consequence, e.g., between 15–21 April 15 to 90% of the flight routes were cancelled implying also significant economic impacts (see, e.g., [1,2]). According to radar and satellite measurements the plumes from Eyjafjallajökull often reached the height of 5–10 km between 14–18 April and 3–20 May [3,4]. The ash particles and gases injected high into the atmosphere were transported mostly by westerly and northwesterly winds towards Europe, and small particles often travelled thousands of kilometers before being removed from the atmosphere. Ash plumes could be detected from several parts of Europe, including Great Britain, Germany, Poland, the Netherlands and Norway, at 1 to 7 km altitude in plumes of 100 m to 3 km depth and 100 to 300 km width [5]. At the beginning of May, due to the northerly flows in the Atlantic region, the ash plumes reached even the Iberian Peninsula within three to five days at an altitude as high as 11–12 km [6,7], and volcanic plumes from the Eyjafjallajökull eruptions were detected as far as Western Siberia, Russia, about 5000 km away from Iceland, on 20–26 April [8,9]. One year later, in May 2011, the volcanic ash clouds of the Grímsvötn volcano (Iceland) quickly rose to 20–25 km in altitude [10] and reached some part of Greenland and Scandinavia within a few days [11,12], impacting the air traffic in Northern Europe. Furthermore, traces of the volcanic clouds could also be detected in the stratosphere over Western Siberia [9]. In March–April 2011, due to the Fukushima Daiichi nuclear disaster (Japan), radioactive materials were transported in the atmosphere over the Pacific Ocean [13–15] causing measurable concentration even in Europe [16–18] within a few weeks in several countries like Greece [19], Germany [20], and Serbia [21]. In April 2015 the plumes from the Calbuco volcano in Chile rising to 15–23 km height [22,23] reached Argentina and Uruguay [24] and influenced even the development of the Antarctic ozone hole in 2015 [25]. In the same year, in Europe, the intense eruption of Mount Etna [26,27] attracted people's attention in December, and its SO_2 plumes circumnavigated the whole Northern Hemisphere in an increasingly stretched filament shape [28].

Massive eruptions with plumes injected into the stratosphere can even have an impact on the global climate. It is the consequence of the ability of small aerosol particles or gases to travel in the atmosphere for a long time—even months or years—before they are removed (see, e.g., [29,30]). Additionally, within this time-frame they become substantially mixed over the hemispheres [31]. For example, the global surface temperature dropped by 0.5–0.7 °C for about two years due to a significant reduction of irradiation as a result of the Mount Pinatubo volcano's eruption in 1991 [32–34].

The rapid spread of pollutants in the atmosphere is due to the fact that in 3-D flows, as is the case for the atmosphere, individual particles carry out a so-called chaotic motion [35,36]. Its typical characteristics are (i) the sensitivity to the initial conditions, which implies that initially nearby particle trajectories diverge rapidly, namely, exponentially within a short time, (ii) the particle's motion is irregular, and (iii) the development of complicated but well-organized fractal structures. The chaotic nature implies that initially small and compact pollutant clouds stretch rapidly in time and evolve into a more and more complicated filamentary and tortuous structure (as can also be seen, e.g., on satellite observations and in model simulations in [12,28]). The intensity of the chaoticity of the pollutant spreading can be studied by means of different quantities. One of them is topological entropy [35,37,38], which, in the atmospheric context, characterizes the rate of the stretching of the length of pollutant clouds distorted into filament-like shapes. Topological entropy is also closely related to the unpredictability of the spreading and the complexity of the structure of a pollutant cloud [30,31,39].

Due to the impact of gravity, aerosol particles move downwards on average, hence they can travel in the atmosphere exhibiting the above-mentioned chaotic behavior only for a finite time interval before they are deposited on the ground. This kind of chaos is called transient chaos [37,38]. It can be shown that the time dependence of the number of non-deposited particles starts to decay approximately exponentially after a while. The rate of this exponential decrease is called the escape rate [29,37,38].

Even though, as the above examples demonstrate, volcano eruptions and industrial accidents may have an impact on a continental and global scale far away from their initial location, many of the students and non-experts are not familiar with the above-mentioned main properties of large-scale atmospheric pollutant spreading and deposition. For example, a common misconception is that pollutant clouds disperse in the atmosphere like dye blobs on clothes. There are some freely available atmospheric dispersion models with which simulations can be carried out, such as the Hybrid Single-Particle Lagrangian Integrated Trajectory (HYSPLIT) model [40,41] which has also a web based user interface, or the Lagrangian analysis tool LAGRANTO [42,43]. Nevertheless, the available models are principally designed for researchers, providing several options for dispersion calculations, and they often do not have user-friendly graphical user interface, as it is the case, e.g., for the FLEXible PARTicle (FLEXPART) dispersion model [44] and for the FALL3D [45–47]. To our knowledge, none of these models are designed to investigate the chaotic features of atmospheric spreading. Therefore, in this study we introduce a Lagrangian model called the Real Particle Lagrangian Trajectory model– Chaos version (RePLaT-Chaos), which specifically aims to demonstrate the chaotic behavior of pollutants. It is freely downloadable from [48]. Due to its easy-to-understand graphical user interface, it is also a suitable tool for students and for other non-experts who are interested in atmospheric spreading phenomena and would like to study this in an interactive way by monitoring the spreading process on maps. Similar educational tools on environmental topics which affect our everyday life have become popular nowadays. For desktop applications which allows students to explore the subject of climate change, see, e.g., Educational Global Climate Model (EdGCM) [49] or Planet Simulator (PlaSim) [50]).

The paper is organized as follows. In Section 2 the two chaotic quantities, the topological entropy and the escape rate, which can be determined by means of the RePLaT-Chaos application, are introduced. Section 3 presents the equations of motions for trajectory calculations, the computation of the topological entropy and escape rate, and a brief overview of the RePLaT-Chaos application. Section 4 demonstrates the applicability and possibilities of RePLaT-Chaos on different examples. It includes a simulation of the spreading of a volcanic ash cloud emanated from the Eyjafjalljökull volcano's eruption. Furthermore, case studies regarding the topological entropy and escape rate are also presented in order to get an impression about their meaning and their magnitudes in different cases. Section 5 summarizes the chaotic characteristics of atmospheric pollutant spreading observable using RePLaT-Chaos and the main features of the application. Appendix A provides a detailed manual for the RePLaT-Chaos application, an overview of the user interface, including the description of its pages, and presents the options for starting new or loading saved simulations. It also contains instructions on how to obtain the topological entropy and the escape rate by means of RePLaT-Chaos.

2. Chaotic Quantities

2.1. Topological Entropy

In dynamical systems theory, topological entropy is a measure of the complexity of the motion [35,36]. Besides its abstract interpretations, its property which is the easiest to capture in measurements is that it also represents the growth rate of the length of line segments. The existence of the topological entropy is a basic property of chaos. A possible definition of chaos is that "a system is chaotic if its topological entropy is positive" [35,36].

As is mentioned in the Introduction, due to the chaotic nature of spreading, pollutant clouds stretch rapidly in time. The growth of the length L of a pollutant cloud in time t is approximately exponential after some days, i.e.,

$$L(t) \sim \exp(ht). \tag{1}$$

Here h is called the topological entropy [35,36,51–53], or the stretching rate in the atmospheric context [30,39]. Topological entropy is the rate of the exponential increase of the filament length. It is a measure of chaoticity, i.e., it quantifies the complexity and irregularity of the advection of a pollutant cloud: the larger the topological entropy, the more quickly the pollutant cloud stretches, the more complicated the shape in which it develops, the more foldings and meanders it contains, and the larger the geographical area the pollutant cloud covers.

2.2. Escape Rate

Transient chaos means that chaotic behavior takes place only for a finite duration. This is the case for the spreading of aerosol particles in the atmosphere [35,37,38]. In this kind of systems, there exists a time-dependent set, the so-called chaotic saddle, which is responsible for the chaotic motion. The trajectories initialized on this saddle would never leave the saddle and carry out chaotic motion for an infinite amount of time. The chaotic saddle is a zero-measure set with fractal structure. As a consequence, in computational simulations using random initial conditions the probability for an initial condition to be located exactly on the chaotic saddle is zero and the trajectories sooner or later leave (i.e., "escape") any arbitrary pre-selected region of the saddle. In the context of the atmospheric pollutant spreading problem, this region can be chosen as the entire atmosphere, therefore, escaping means the deposition of particles. After a sufficiently long time t_0, the decay in the ratio $n(t)/n(0)$ of survivor particles is approximately exponential in transiently chaotic systems:

$$n(t)/n(0) \sim \exp(-\kappa t) \text{ for } t > t_0. \tag{2}$$

The coefficient κ is called the escape rate [35,37,38] and in the case of pollutant spreading it characterizes the speed of the deposition. Larger escape rate implies faster deposition process, i.e., more particles leaving the atmosphere up to a given time instant. In general, the average lifetime of the particles after t_0 can be estimated by κ^{-1}.

3. Methods

RePLaT-Chaos is a simpler version of the previously developed Real Particle Lagrangian Trajectory (RePLaT) model [29,54,55]. It computes the trajectories of individual spherical particles of realistic size and density, taking into account advection and the role of gravity through the terminal velocity of individual particles. In this sense, RePLaT-Chaos (and RePLaT) differ from the dispersion models which track so-called computational particles, like FLEXPART [44] and HYSPLIT [40,41], i.e., when each particle carries a certain amount of mass assigned to them upon the release, and this mass can be changed, e.g., due to deposition processes. In contrast to this, in RePLaT-Chaos, each particle has its own radius and density (and thus, its own realistic mass), and the effect of gravitational settling is calculated individually for each particle based on its own properties. Consequently, if a particle deposits on the surface, the entire particle remains there, not only a certain ratio of its mass. This individual particle approach is essential in order for the chaotic features of spreading to be studied appropriately. A pollutant cloud in the simulations consists of such kind of particles.

The computational background and the validity of RePLaT-Chaos, the RePLaT model, was tested in a number of cases. By simulating the spreading of volcanic ash injected in the atmosphere during the eruption of the Eyjafjallajökull and Mount Merapi [29,56] a reasonable agreement was found between the distribution of volcanic ash in the simulations and in the satellite observations at different time instances over days. Furthermore, the simulation of the spreading and deposition of radioactive materials continuously released during the Fukushima Daiichi nuclear power plant disaster showed that the arrival times of the pollution at different remote locations (e.g., Chapel Hill, Richland (USA), Stockhom (Sweden)) coincided with the measurements, and the RePLaT simulations were able to reproduce even the measured concentrations with acceptable accuracy [57].

3.1. Calculation of Particle Trajectories

For small and heavy aerosol particles it can be shown that a particle is advected by the wind components in the horizontal direction and their vertical motion is influenced by its terminal velocity and the vertical velocity component of air (see, e.g., [29]). RePLaT-Chaos utilizes meteorological data given on a regular longitudinal–latitudinal grid horizontally and at different pressure levels vertically. Therefore, the equations of motion of the particles are written in spherical coordinates in the horizontal direction and in pressure coordinates in the vertical direction in agreement with the structure of the meteorological data:

$$\frac{d\lambda_{\mathrm{P}}}{dt} = \frac{u(\lambda_{\mathrm{P}}(t), \varphi_{\mathrm{P}}(t), p_{\mathrm{P}}(t), t)}{R_E \cos \varphi_{\mathrm{P}}} = u_{\mathrm{rad}}(\lambda_{\mathrm{P}}(t), \varphi_{\mathrm{P}}(t), p_{\mathrm{P}}(t), t), \tag{3}$$

$$\frac{d\varphi_{\mathrm{P}}}{dt} = \frac{v(\lambda_{\mathrm{P}}(t), \varphi_{\mathrm{P}}(t), p_{\mathrm{P}}(t), t)}{R_E} = v_{\mathrm{rad}}(\lambda_{\mathrm{P}}(t), \varphi_{\mathrm{P}}(t), p_{\mathrm{P}}(t), t), \tag{4}$$

$$\frac{dp_{\mathrm{P}}}{dt} = \omega(\lambda_{\mathrm{P}}(t), \varphi_{\mathrm{P}}(t), p_{\mathrm{P}}(t), t) + \omega_{\mathrm{term}}(\lambda_{\mathrm{P}}(t), \varphi_{\mathrm{P}}(t), p_{\mathrm{P}}(t)) \tag{5}$$

where λ_{p} and φ_{p} are the longitude and latitude coordinates, $p_{\mathrm{p}}(t) \equiv p(\lambda_{\mathrm{P}}(t), \varphi_{\mathrm{P}}(t), p_{\mathrm{P}}(t), t)$ is the pressure coordinate of a particle, $R_E = 6370$ km is the Earth's radius, u and v are the zonal and meridional velocity component of the air in the units of $\mathrm{m\,s^{-1}}$, u_{rad} and v_{rad} are the same but in units of $\mathrm{s^{-1}}$ fitted to the longitude–latitude coordinates, ω is the vertical air velocity component in the pressure system, and ω_{term} is the corresponding terminal velocity of the particle in motionless air of the form of

$$
\omega_{\mathrm{term}} = \begin{cases} \dfrac{2}{9} r^2 \dfrac{\rho_{\mathrm{p}}}{\nu} g^2, & \text{if } \mathrm{Re} \ll 1 \\[2ex] \sqrt{\dfrac{8}{3} \dfrac{\rho_{\mathrm{p}}\rho r}{C_{\mathrm{D}}} g^3}, & \text{if } \mathrm{Re} \gg 1. \end{cases}
\tag{6}
$$

Here r and ρ_{p} are the radius and the density of the particle, respectively, ν and $\rho = p/R_{\mathrm{d}}/T$ are the kinematic viscosity and density of air, respectively, $R_{\mathrm{d}} = 287\,\mathrm{J\,kg^{-1}\,K^{-1}}$ is the specific gas constant of dry air, g denotes the gravitational acceleration, C_{D} is the drag coefficient (for particles assumed to be spheres $C_{\mathrm{D}} = 0.4$), and $\mathrm{Re} = 2rV/\nu$ is the Reynolds number (where V is the instantaneous particle velocity). The limit of $r = 0\,\mu\mathrm{m}$ can be considered as gas "particles", the terminal velocity of which is $\omega_{\mathrm{term}} = 0$.

The dependence of kinematic viscosity ν on temperature T and pressure p is calculated according to Sutherland's law [58]

$$
\nu = \beta_0 \frac{T^{3/2}}{T + T_{\mathrm{S}}} \frac{R_{\mathrm{d}} T}{p}.
\tag{7}
$$

Here $\beta_0 = 1.458 \times 10^{-6}\,\mathrm{kg\,m^{-1}\,s^{-1}\,K^{-1/2}}$ is Sutherland's constant and $T_{\mathrm{S}} = 110.4\,\mathrm{K}$ is a reference temperature.

RePLaT-Chaos solves the differential Equations (3)–(5) by an explicit second-order Runge–Kutta method, i.e., by the second-order Petterssen scheme (applied often also in other Lagrangian dispersion models [40,41,44]). Hence the position $\mathbf{r}(t + \Delta t) = [\lambda(t + \Delta t), \varphi(t + \Delta t), p(t + \Delta t)]$ of a particle at time instant $t + \Delta t$, using the velocity $\mathbf{v}(\mathbf{r}(t), t) = [u_{\mathrm{rad}}(\mathbf{r}(t), t), v_{\mathrm{rad}}(\mathbf{r}(t), t), \omega(\mathbf{r}(t), t) + \omega_{\mathrm{term}}(\mathbf{r}(t), t)]$ at time t, reads as:

$$
\mathbf{r}_0(t + \Delta t) = \mathbf{r}(t) + \mathbf{v}(\mathbf{r}(t), t)\Delta t,
\tag{8}
$$

$$
\mathbf{r}(t + \Delta t) = \mathbf{r}(t) + \frac{1}{2}\left(\mathbf{v}(\mathbf{r}(t), t) + \mathbf{v}(\,\mathbf{r}_0(t + \Delta t), t + \Delta t)\right)\Delta t.
\tag{9}
$$

The utilized meteorological data should be available on a regular latitude–longitude grid on different pressure levels with a given (e.g., 3 or 6 h) time resolution. Therefore, in order to solve the equations of motion of the particles and to calculate the particle trajectories, the quantities u, v, ω, T are interpolated to the location of the particles in each time step. RePLaT-Chaos applies linear interpolation in each of the three directions and in time.

Users have the option to choose between variable time step and constant time step for the trajectory calculation. In the former option, the maximum time step Δt_C for each particle is determined based on the grid size and the current atmospheric velocity components as

$$
\Delta t_C = C \min\left\{ \frac{\Delta\lambda_{\mathrm{g}}}{|u_{\mathrm{rad}}(\mathbf{r}(t), t)|}; \frac{\Delta\varphi_{\mathrm{g}}}{|v_{\mathrm{rad}}(\mathbf{r}(t), t)|}; \frac{\Delta p_{\mathrm{g}}}{|\omega(\mathbf{r}(t), t) + \omega_{\mathrm{term}}(\mathbf{r}(t), t)|} \right\}
\tag{10}
$$

with $C = 0.2$ where $\Delta\lambda_{\mathrm{g}}$ [rad], $\Delta\varphi_{\mathrm{g}}$ [rad] and Δp_{g} [Pa] denote the grid size in longitudinal, meridional and vertical direction, respectively. By means of such a choice the smallest features resolved by the meteorological fields are taken into account as pointed out in [59]. The minimal time step $\Delta t_{\min}$ is determined by the user, therefore, $\Delta t = \max\{\Delta t_C; \Delta t_{\min}\}$. If the obtained time step Δt would be larger than the time interval $(t_{\mathrm{next}} - t)$ up to the next writing of particle data to file or up to the next reading of new meteorological fields, then the time step is modified as

$$
\Delta t = \min\{t_{\mathrm{next}} - t, \max(\Delta t_C; \Delta t_{\min})\}.
\tag{11}
$$

3.2. Calculation of the Topological Entropy

Topological entropy is calculated by RePLaT-Chaos as in [30,39]. In order for the length of a pollutant cloud to be appropriately determined, the user should initiate 1-D "pollutant clouds", i.e., line segments or filaments. The length of a filament is the sum of the distances of its neighboring particle pairs:

$$L(t) = \sum_{i=1}^{n(t)-1} |\mathbf{r}_i(t) - \mathbf{r}_{i+1}(t)|, \tag{12}$$

where $\mathbf{r}_i$ is the position of the ith particle and $n(t)$ is the number of particles. The distance $|\mathbf{r}_i(t) - \mathbf{r}_{i+1}(t)|$ in units of km are calculated along great circles neglecting the vertical stretching which proved to be 10^{-2} to 10^{-3} times smaller than the horizontal one [31]:

$$|\mathbf{r}_i(t) - \mathbf{r}_{i+1}(t)| = \arccos\left[\sin \varphi_i \sin \varphi_{i+1} + \cos \varphi_i \cos \varphi_{i+1} \cos(\lambda_i - \lambda_{i+1})\right] \times \frac{180}{\pi} \times 111.1, \tag{13}$$

where λ_i and φ_i are the zonal and meridional coordinate of the ith particle, respectively. The factor $\frac{180}{\pi} \times 111.1$ converts the unit from radian to kilometer using the fact that the spherical distance of $1°$ along a great circle corresponds to a length of 111.1 km along the surface. Note that a filament remains a single filament forever, and cannot split up into two or more branches, because it would require a wind vector that points in more than one direction at some location. Hence the determination of the full (folded) length is unambiguous.

Since subsequent particles may travel far away from each other in time, the length of a pollutant cloud that consists of a finite number of particles may be underestimated compared to a pollutant cloud with the same initial condition consisting of an infinite number of particles (i.e., "continuous" pollutant cloud). This implies that after a certain amount of time, when there are several "cut-off" segments among the particles, the computed length differs from the expected approximately exponential function as its values are lower than the expected ones. Therefore, in order to reduce the underestimation of the length, there is an option for users that if the distance of two neighboring particles becomes larger than a threshold distance defined by the user, a sufficient number of new particles is inserted between them uniformly.

Based on Equation (1) the topological entropy is determined as the slope h of a linear least squares fit applied to the natural logarithm of the length $L(t)$ of the filament for the time interval chosen by the user.

3.3. Calculation of the Escape Rate

In order to determine the escape rate it is worth tracking the trajectories of a large number of particles until they leave the atmosphere one by one. At each time instant t RePLaT-Chaos determines the number $n(t)$ of the particles still moving in the atmosphere, i.e., the number of the non-escaped particles. Based on Equation (2) the value of the escape rate κ is calculated as $(-1) \times$ the slope of a linear least squares fit applied to the natural logarithm of the ratio $n(t)/n(0)$ of non-escaped particles [29,55] for the time interval chosen by the user. It is worth noting that the time t_0 in Equation (2), after which the exponential decay starts, depends on the initial conditions, the initialization time instant and the properties of the particles as well.

3.4. RePLaT-Chaos in a Nutshell

RePLaT-Chaos is a desktop application with user-friendly graphical user interface and simulates the atmospheric spreading of pollutant clouds in the time interval and with simulation setups given by the user. Pollutant clouds consist of individual particles, the number of which is determined by the user. The initial position, size, and other properties of the pollutant cloud (and its particles) can be set up on the user interface, and pre-generated pollutant clouds can also be read for the simulations. For the spreading calculations, meteorological files containing the appropriate meteorological data that overlap the defined time interval are required. RePLaT-Chaos determines the new position of each particle of the pollutant cloud from Equations (3)–(5) in each time step based on the meteorological data and writes the particle data to file. Furthermore, there are options for computing the length of the pollutant cloud or the ratio of the particles not deposited from the atmosphere. These data are needed for the calculation of the two quantities characterizing the chaoticity of the spreading: the topological entropy and the escape rate. RePLaT-Chaos provides an opportunity to replay simulations saved in files and to determine the above-mentioned two chaotic measures. The detailed manual for the application can be found in Appendix A. In the next section, the applicability of RePLaT-Chaos are presented, drawing attention to the main features of the large-scale atmospheric spreading of pollutants and its chaotic characteristics.

4. Results from RePLaT-Chaos Simulations

4.1. Spreading of a Volcanic Ash Cloud Emitted during the Eyjafjalljökull Volcano's Eruption

The Eyjafjalljökull volcano in Iceland showed an increased seismic activity in the spring of 2010. After the first eruption on 20 March, one of the most intense eruptions happened on 14 April 2010 [60]. For about four days, the vertical extent of the emitted ash columns often exceeded the height of 4 to 5 km, with the top of the column occasionally reaching even the altitude of 10 km according to weather radar, LIDAR, and satellite measurements (see, e.g., [3,4]). The mean size and density of the particles which travelled across Europe were found to be between $r \approx 0.1$ to 10 µm and $\varrho_p \approx 2000$ kg m^{-3}, respectively (see, e.g., [5,61–63]).

Based on these data, to get a first impression about the main characteristics of atmospheric pollutant spreading by means of RePLaT-Chaos, Figure 1 shows the simulation of the spreading of a single, initially compact ash cloud of height of 4 km injected into the atmosphere due to the eruption of the Eyjafjallajökull on 14 April 2010 at 06:00 UTC. The ash cloud in the simulation consisted of 2.7×10^4 particles with $r = 5$ µm and $\varrho_p = 2000$ kg m^{-3}.

Figure 1 illustrates that within a few days the ash cloud travels over Scandinavia and reaches Eastern Europe due to being transported by the northwesterly winds of a high pressure system located south of Iceland at the beginning and then moving towards Scandinavia. Figure 1 demonstrates well that the spreading of volcanic ash clouds (and any atmospheric pollutants) differs from the dispersion of dye droplets on clothes. The latter is of a slowly growing circular shape, while Figure 1 shows that an important feature of atmospheric pollutant spreading is the rapid distortion of an initially small and compact cloud into an increasingly stretched, filament-like shape, extending to a region of some thousands of kilometers within a few days. As mentioned in the Introduction, the observed rapid stretching of pollutant clouds is a consequence of the chaotic nature of the spreading. Therefore, the rate of the stretching is a possible measure of the strength of chaos which will be illustrated through some examples in Section 4.2.

At the beginning (see Figure 1a), the top of the ash cloud reaches the altitude of 9 km (cyan color). However, due to the impact of gravity, the particles descend in the atmosphere more or less continuously (but not uniformly), and after two days they reach the altitude of about 4–6 km (green color, Figure 1c). Within three days, the altitude of the ash cloud in an extended region decreases even below 2–3 km (yellow color, Figure 1d). After 10 days a large number of particles are found to be deposited on the ground (black color, Figure 1f) across Siberia. The deposition distribution shows another important characteristic, typical of chaotic phenomena, namely that it is inhomogeneous with filamentary structure, with denser and sparser regions. Additionally, it can be also seen that particles do not fall out from the atmosphere at almost the same time as a coherent patch but rather some parts of the ash cloud are deposited by the 7th day after the eruption already (Figure 1e), while several particles are still in the middle of the troposphere, at an altitude of about 5 km (green) even after 10 days (Figure 1f). As it is introduced in Section 1, this kind of deposition dynamics is characteristic to transiently chaotic phenomena. The measure of the rapidity of deposition processes will be discussed in Section 4.3 in detail.

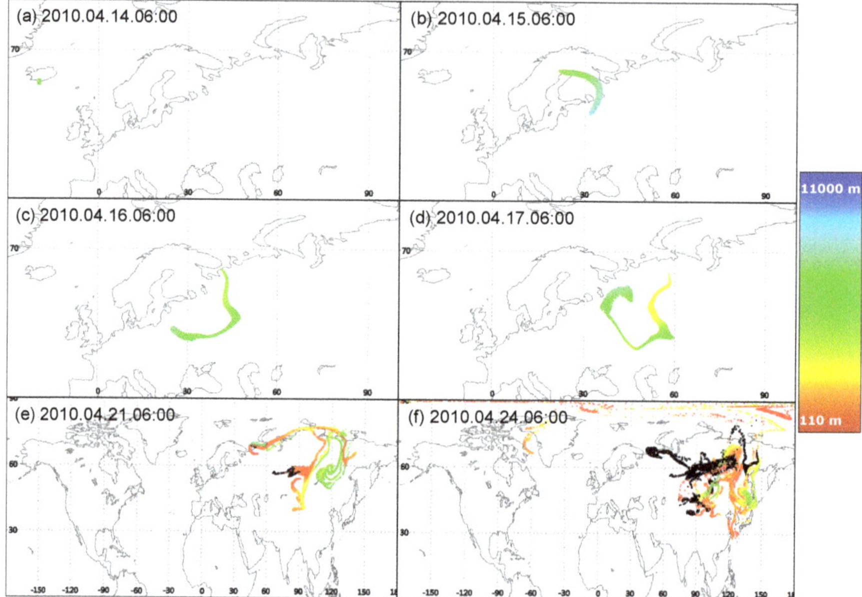

Figure 1. Simulation of the spreading of volcanic ash particles from the Eyjafjallajökull volcano's eruption at different time instants in the form of "year.month.day.hour:minute" indicated in the panels' label (**a–f**). $30 \times 30 \times 30$ particles of $r = 5$ μm and $\varrho_p = 2000$ kg m^{-3} are initiated in a rectangular cuboid of size of 100 km $\times$ 100 km $\times$ 4 km at 63.63° N, 19.6° W, at the altitude of 7 km on 14 April 2010 at 06 UTC. Simulation is initialized with the parameters on the left of Figure A1. Colorbar indicates the altitude of the particles, black color marks deposited particles.

4.2. Stretching of the Pollutant Clouds—The Topological Entropy

Section 4.1 has shown that even an initially cuboid-shaped pollutant cloud soon becomes distorted into a tortuous, filamentary shape due to the chaotic nature of atmospheric spreading, and the extension of the cloud grows rapidly. To quantify this growth, by means of RePLaT-Chaos application, the time-dependence of the length increase of 1-D pollutant clouds (i.e., lines or filaments) can be measured. The stretching rate of the length, the topological entropy h in Equation (1), quantifies the intensity of the underlying chaotic dynamics which the pollutant cloud is subjected to during spreading.

To get an impression of the meaning and consequences of the value of the topological entropy h, Figure 2 illustrates the distribution of two meridional line segments (having the same length at the emission) after 10 days and the corresponding curves of their length increase. Both cases show that the length of the filaments indeed grows in an approximately exponential manner in time (Figure 2b,d) after a few days (as a line in the semi-logarithmic plot). In Figure 2b the slope of the linear fit is $h = 0.808$ day^{-1}, while in Figure 2d the slope is found to be about 56% smaller, $h = 0.357$ day^{-1}. These values mean that in every $h^{-1} = 1.238$ and 2.801 days the length of the pollutant cloud stretches by a factor of e ≈ 2.718, respectively. With h being in the exponent in Equation (1), this approximately double factor between the topological entropies results in the fact that the length of the filaments after 10 days is about 1.242×10^6 km for the filament initiated in Europe and 9.660×10^3 km for the one emitted in Africa (calculated as $\exp(14.032)$ and $\exp(9.176)$, respectively, reading the length data on April 24 at 6 UTC from the graphs.). The nearly 100-fold difference in their length (and the corresponding deviation of their topological entropies) obviously implies remarkably different distribution patterns at the end of the simulation. While $h = 0.357$ day^{-1} in Figure 2c is associated with a slightly crumpled filament which has not travelled far away from its initial location as it is drifting slowly with the trade winds near the Equatorial region, the filament in Figure 2a has a much more complicated shape with several foldings and meanders that cover a considerable part of the Northern Hemisphere. We note that, in general, larger topological entropy values and more intense spreading characterize the pollutant clouds initiated at the mid- and high latitudes than the ones start in the tropical region [30,31]. This is a consequence of the enhanced cyclonic activity in the extratropics associated with intensified shearing and mixing effects on the pollutant clouds. It is also worth noting that certain atmospheric features can be identified based on the pattern formed by the particles of the pollutant clouds: e.g., in Figure 2a south of Greenland and east of Scandinavia the trace of two cyclones can be noticed drawn by the spiral formations of the pollutant cloud.

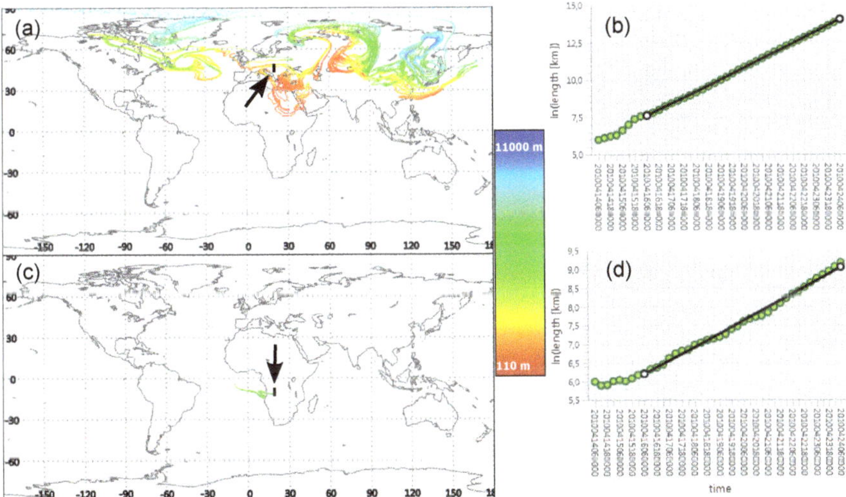

Figure 2. (**a**,**c**) The advection pattern of a pollutant cloud after 10 days, and (**b**,**d**) the time dependence of the logarithm of the length of the pollutant clouds, respectively. The simulations are initialized with the simulation parameters on the left of Figure A1 but with top and bottom reflection coefficients of 1 and inserting new particles if the distance of two particles became greater than 100 km. The pollutant clouds are initialized as meridional line segments of 400 km at the altitude of 5500 m at (**a**,**b**) 47° N, 19° E and (**c**,**d**) 10° S, 19° E. They consist of 1000 particles at the beginning. Their initial position is indicated by the thick black lines to which arrows are pointing. The particle radius is 0 μm. Colorbar indicates the altitude of the particles. In panel (**b**,**d**) the black line indicates a linear fit to the logarithm of the length for the time interval from 16 April, 6 UTC to 24 April, 6 UTC. Its slope is (**b**) $h = 0.808$ day^{-1} and (**d**) $h = 0.357$ day^{-1}.

4.3. Deposition of the Particles—The Escape Rate

Section 4.1 draws attention to the fact that the lifetime of particles even in an initially small pollutant cloud may be quite different. The reason behind the observed differences is that the particles do not fall directly purely vertically from their initial position onto the ground, but travel along complicated trajectories due to the chaotic nature of spreading. In this way their vertical movement is affected by both their terminal velocity and the local instantaneous vertical component of the air. Both the terminal velocity ω_{term} (Equation (6)) through kinematic viscosity ν and/or air density ϱ and the vertical air velocity v (Equation (5)) depend on the position of the particle and the time instant. For light aerosol particles the value of the upward directional vertical air velocity often exceeds the downward effect of their terminal velocity, thus, besides falling downwards on average, these particles have more chance to move also upwards in the atmosphere with the flows. The chaotic nature of spreading implies that nearby particles may reach remote locations within short times where they are also subjected to different vertical velocities, therefore, they may be deposited at considerably different time instants and locations.

In order to study the process of deposition, Figure 3a,b shows how the ratio of non-deposited particles initially distributed uniformly over the globe at the altitude of 5.5 km changes in time. At the beginning of the simulation, a short plateau can be seen for both particles of radius $r = 7$ μm (Figure 3a) and $r = 9$ μm (Figure 3b), which indicates that a certain time is needed even for the "fastest falling" particles to reach the surface. After a short transient, the plateau is followed by an approximately exponential decrease in the ratio of non-escaped particles, the rapidity of which is characterized by the escape rate κ in Equation (2). The escape rate is smaller ($\kappa = 0.278$ day^{-1}) for smaller particles (Figure 3a) and larger ($\kappa = 0.489$ day^{-1}) for larger particles (Figure 3b), as expected naively, but it depends on the atmospheric conditions, too. In fact, the dependence of κ on r for $r \leq 10$ μm particles proved to be quadratic in a recent research studying aerosol particles with a realistic density of 2000 kg m^{-3} [55]. This is in harmony with the fact that the updrafts and downdrafts in the atmosphere approximately balance each other's effect on the particles, thus particles in rough average fall with their terminal velocity ω_{term}, which depended quadratically on r for these small particles with Re $\ll 1$ (Equation (6)). The obtained κs imply that after the exponential decay takes place, after $\kappa^{-1} = 3.597$ and 2.045 days, only a proportion of e$^{-1} \approx 0.368$ of the particles can still be found in the air, respectively. It is worth noting that the reciprocal of κ is often considered to be a rough estimate of the average lifetime of typical particles in the exponentially decreasing stage [37,38].

Figure 3a,b also confirms an interesting observation made in Section 4.1 that even identical particles may often have significantly different lifetimes. For example, the first particles in Figure 3a leave the atmosphere on 15 April, only one day after the emission, while after more than two weeks, on 30 April, there are still 1.4% of the particles (exp(-4.256) from the data of the graph) drifting in the atmosphere. Simulating the atmospheric spreading of a larger number of particles with $r = 7$ μm for a longer time period, it turns out that a ratio of 10^{-5}–10^{-6} of the particles is able to survive more than two months in the atmosphere, as well as that the initial location of the long- and short-living particles folds into each other in thin filaments in a fractal structure in extended regions [55].

Figure 3c demonstrates that the inhomogeneity and irregularity in the pattern of the deposited particles in Figure 1f is not the consequence of the initially small extension of the volcanic ash cloud studied in Section 4.1. The filamentary deposition pattern with denser and sparser regions, typical for transient chaos, can also be seen even for particles initially distributed completely uniformly over the whole globe.

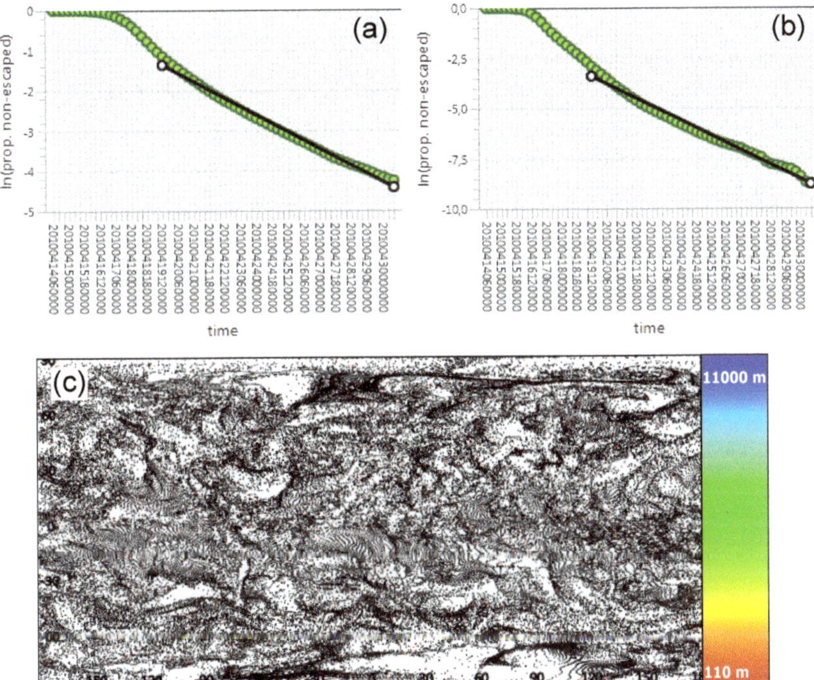

Figure 3. (**a**,**b**) The time-dependence of the logarithm of the ratio of non-escaped particles. The simulations are initialized with the simulation parameters on the left of Figure A1 but with an end date of "2010.04.30.12:00:00" and with calculating the ratio of non-escaped particles. The pollutant clouds are initialized as $300 \times 300 \times 1$ particles at $0°$ N, $0°$ E and at the altitude of 5500 m with an extension of 4×10^4 km $\times\ 2 \times 10^4$ km $\times\ 0$ m (i.e., covering the entire globe uniformly at a single altitude). The particle density is 2000 kg m^{-3} and the particle radius is (**a**) 7 µm and (**b**) 9 µm, respectively. The black lines indicate linear fits to the logarithm of ratio of non-escaped particles for the time interval from 19 April, 12 UTC to 30 April, 12 UTC. The $(-1) \times$ slopes are (**a**) $\kappa = 0.278$ day^{-1} and (**b**) $\kappa = 0.489$ day^{-1}. (**c**) The deposition pattern of the particles in panel (**b**) at 30 April 2010 at 12 UTC. Colorbar indicates the altitude of the particles, black color marks deposited particles.

5. Conclusions

In this paper, the Lagrangian particle-tracking trajectory model RePLaT-Chaos is introduced and is shown to be applicable for the study of the main features of atmospheric pollutant spreading, which are also discussed in detail. Due to its user-friendly graphical user interface, RePLaT-Chaos is a suitable tool for anyone who is interested in studying the characteristics of the atmospheric spreading of pollutants. Users can design their own "volcano eruptions" changing the location, altitude, extent of the pollutant clouds, as well as the number of the tracked particles and their density and diameter. It can be easily observed how these parameters alter the spreading, and other interesting questions can also be studied, e.g.:

- How much faster do particles with larger size/higher density leave the atmosphere compared to smaller/lighter ones?
- Do the particles deposit on the surface in the shape of patches or in a filamentary structure?
- Is it possible for the particles of an initially small and compact pollutant cloud to cover more or less homogeneously the hemisphere where they are initialized, or the whole globe, and how long does it take?

- How does the initial geographical location of the pollutant clouds affect the rate of their stretching?
- Does the rapidity of the deposition or the stretching of a pollutant cloud depend on initialization time, e.g., the season in a year?
- Can cyclones, jet streams, etc. be revealed by tracking pollutant particles?

By means of RePLaT-Chaos, it can be easily shown that the spreading of volcanic ash and other atmospheric pollutants is peculiar, because it is an example of what is called a chaotic process. One can reveal that the basic difference between the dispersion of a dye droplet on clothes and the spreading of volcanic ash in the atmosphere is that the former grows slowly in a compact shape, while the latter becomes rapidly distorted into a filament, the length of which increases quickly in time. Furthermore, users can become acquainted with the basic concepts of chaos on their own. They experience the rapid divergence of nearby trajectories, the particles' irregular motion in the atmosphere, and the above-mentioned quick development of pollutant clouds into a filamentary, tortuous and complicated but yet organized shape with many foldings and meanders.

Users can easily assign two quantities to their spreading events to characterize the chaotic behavior. One of them is the stretching rate of the pollutant clouds, the topological entropy: the greater its value the more quickly the length of the pollutant cloud grows, and the more foldings and complicated shape it has. Therefore, it can be considered as the measure of the strength of chaos and of the unpredictability of the spreading. The other eligible quantity, the escape rate, describes the rapidity of the approximately exponentially decaying process of particle deposition. Based on the graphs of the non-deposited particles, the users can observe on their own the quite different lifetimes of even identical aerosol particles injected into the atmosphere at very nearby geographic locations at the same time instant. In this way, RePLaT-Chaos can be considered as an educational reconstruction of results obtained from contemporary research regarding atmospheric spreading of pollutant clouds and chaotic advection.

As an outlook, we mention that RePLaT-Chaos has another version called RePLaT-Chaos-edu [64] with the same computational background and fewer simulation parameter options but with a more colorful user interface, designed especially for secondary school students. It is intended to serve as a tutorial about the main features of atmospheric pollutant spreading phenomenon. Therefore, besides allowing students to design their own pollution events, it includes a lot of eye-catching animations and easy-to-understand explanations in order to draw the students' attention to the phenomena. The software was tested with a few group of students and received positive feedback [65].

Funding: This research was funded by the János Bolyai Research Scholarship of the Hungarian Academy of Sciences and by the National Research, Development and Innovation Office—NKFIH under grants PD-121305, PD-132709, FK-124256 and K-125171.

Acknowledgments: Fruitful discussions with T. Tél on chaotic advection and how to introduce it to students are gratefully acknowledged. The author thanks M. Herein and M. Vincze for testing the application, and G. Drotos, K. Klemm, M. Pinter and I. Takacs for suggestions on wording.

Conflicts of Interest: The author declares no conflict of interest.

Abbreviations

The following abbreviations are used in this manuscript:
RePLaT   Real Particle Lagrangian Trajectory model

Appendix A. RePLaT-Chaos Manual

Appendix A.1. First Steps and Data Format Requirements

Appendix A.1.1. Launching RePLaT-Chaos

RePLaT-Chaos can be downloaded from the website of the Institute for Theoretical Physics, Eötvös Loránd University [48]. Both a Windows executable installer file and a compressed file including a Java Archive application (usable also on Linux platforms) are accessible. In the former case, the installer installs the application in the folder RePLaT-Chaos in a user selected location. RePLaT-Chaos can be launched by clicking RePLaT-Chaos.exe in the folder. In the latter case the downloaded zip file should be unpacked, and the application can be launched, e.g., from the command prompt by typing the java -jar RePLaT-Chaos.jar command from folder RePLaT-Chaos. In both cases the folder RePLaT-Chaos has a sub-folder named default which contains the default values for the text fields of the user interface (default/default_*.txt files) and the continents.txt file for displaying the map for the simulations. Therefore, this folder and its contents should not be renamed or removed, however, the content of the `default_*.txt` files may be changed preserving their formats. Sample meteorological data in the required format are also available on the website for a 16-day time period overlapping with the Eyjafjallajökull volcano's eruptions in 2010.

On the user interface the menu items and buttons with underlined letter/number can be reached not only by mouse clicks but also by keyboard shortcut `Alt + <letter>/<number>`.

Appendix A.1.2. Input Meteorological Data

For the simulation of the spreading of pollutant clouds meteorological data in Network Common Data Form (NetCDF) format files are needed. Data should be downloaded for the entire globe and at least six hours of time resolution. Such data are accessible in different temporal and spatial resolutions, e.g., from the European Center for Medium-Range Weather Forecasts (ECMWF) datasets [66] (ERA-40, ERA-Interim, etc.). The properties of the files required by RePLaT-Chaos are:

- Meteorological data on a regular longitude–latitude grid ($[0° \text{ E} : \Delta\lambda_g : 360° \text{ E} - \Delta\lambda_g] \times [90° \text{ S} : \Delta\varphi_g : 90° \text{ N}]$) at different pressure levels, where $\Delta\lambda_g$ and $\Delta\varphi_g$ are the horizontal grid resolutions. The software considers the lowest level as Earth's surface and the highest one as the "top" of the simulation region.
- Required meteorological variables: zonal [m/s] and meridional [m/s] wind components, vertical velocity component [Pa/s] and temperature [K].
- A meteorological file contains the values of one of the meteorological variables on the above-mentioned grid for a single time instant.
- File name convention: `<variable name><yyyyMMddhhmmss>.nc`, where the format `<yyyyMMddhhmmss>` denotes the following: year (yyyy), month (MM), day (dd), hour (hh), minute (mm), second (ss) given in 4, 2, 2, 2, 2, and 2 digits, respectively.

Appendix A.1.3. Output Data

RePLaT-Chaos writes the data of the particles of the pollutant clouds to comma-separated values text files with CSV file extension, therefore, these files can be easily read or analyzed with other tools, too. An output file represents again a single time instant and contains one line for each particle. The file name convention is `<file name pattern><yyyyMMddhhmmss>.csv`. The comma separated data in a line are: $\lambda, \varphi, z, r, \rho_p, \iota$, where

- λ is the longitude coordinate of the particle [rad] $\in [0, 2\pi)$,
- φ is the latitude coordinate of the particle [rad] $\in [-\pi, \pi]$,
- z is the vertical coordinate of the particle [m] calculated from its pressure coordinate p based on the equations of the standard atmosphere [67],
- r is the particle radius [μm] $\in [0, \infty)$,
- ρ_p is the particle density [kg m^{-1}] $\in [0, \infty)$,
- ι represents whether the particle is in the atmosphere yet [1] or not [0].

The application computes the length of the pollutant cloud [km] and/or the ratio of the non-escaped particles if the user chooses this/these option(s). It writes the natural logarithms of these quantities to file at the time instants given by the user. The file of the length and the ratio of the non-escaped particles contain lines of format of `<yyyyMMddhhmmss><tab>`ln(value of the quantity).

Appendix A.2. Running a Simulation

In RePLaT-Chaos two setup options are available for a new simulation. In the first one parameters both for the simulation and for the pollutant cloud should be given. This screen is accessible via File menu clicking menu item New simulation—set parameters. The other options is New simulation—read parameters. In the latter case, pollutant clouds are not initialized according to user-given parameters rather its particles are read from a file.

Appendix A.2.1. Setting the Simulation Parameters

The user chooses either the New simulation—set parameters or the New simulation—read parameters menu item, in both cases the simulation parameters should be given at first (left panel of the screen in Figures A1 and A2). These parameters are the following:

- Start: start date and time of the simulation (format: <yyyy.MM.dd.hh:mm:ss>).
- End: end date and time of the simulation (format: <yyyy.MM.dd.hh:mm:ss>).
- Time step: if it is constant then the text field represents the value of the constant time step [s] (format: integer); if it is variable then the text field corresponds to Δt_{min} defined in Section 3.1.
- Input folder: folder of the meteorological files. Clicking the Choose input folder button it can be chosen or it can be written directly to the text field.
- File name pattern for the met. fields: the part of the file name before date and time for the meteorological files of the zonal (u), meridional (v), vertical (w) velocity component of air and the temperature (T).
- Reflect from the surface?: if the box is checked each particle bounces back from the lowest meteorological level according to the Reflection coefficient for the surface (format: real). For topological entropy (length) calculation, the box is worth checking and 1 should be given for the reflection coefficient.
- Reflect from the top?: if the box is checked each particle bounces back from the highest meteorological level according to the Reflection coefficient for the top (format: real). Generally, the box is worth checking unless the user especially wants to study how many particles leave the meteorological region at the highest level.
- Save particle data?: should the data of the particles of the pollutant cloud (and the length data and the ratio of the non-escaped particles) be written to file. For example, user should not check the box if he/she carries out test calculations and would like to see the spreading of pollutant cloud only once, i.e., when the user does not need the data later.
- Output folder: the folder of the files for particle data, length data and ratio of the non-escaped particles. Clicking the Choose output folder button it can be chosen or it can be written directly to the text field.
- File name pattern for the output: the part of the particle data file name before date and time.
- Calculate length? (filename): should the length of the pollutant cloud be calculated, and if yes, what should be the name of the length file to which the data are written. The length of the pollutant cloud is computed as the sum of the distances between the subsequent particles as described in Section 3.2. Therefore, the result of the calculation equals to the real length in the unit of km only if the particles are initiated as a one-dimensional "pollutant cloud", i.e, a line segment.
- Calculate ratio of non-escaped particles? (filename): should the ratio of the non-escaped particles be calculated, and if yes, what should be the name of the escape file to which the data are written.
- Time interval for the output: the time interval [s] for writing the data of the pollutant cloud, of the length and of the ratio of the non-escaped particles to file (format: integer)
- Insert new particles if the distance [km] of two particles: should new particles be inserted in the pollutant cloud if the distance of two subsequent particles is greater than the given distance (format: real). The box is worth checking for length calculation; otherwise subsequent particles may travel far away from each other which results in the underestimation of the length of the pollutant cloud as described in Section 3.2.

- Max. number of particles: The number of the particles in the simulation, including the inserted ones, does not exceed the given number (format: integer).

By default, the fields are filled with the default setting parameters loaded from default/ default_simulation_setup.txt. After overwriting any field, the user can reload the default values by clicking the Default button, new data can be loaded in the fields from a chosen file by clicking the Load button, or the values of the fields can be saved in a new file by clicking the Save button. If there is a wrong or empty text field, data could not be saved: the problem is indicated by an alert window.

At first, for starting a simulation the generation of the simulation setup is required: the user can generate it by clicking the Generate simulation setup button. If there is a wrong or empty text field, similarly to saving, the problem is indicated by an alert window. If there is no wrong or empty text field, a pop-up window indicates that the Simulation setup is generated. Then the disabled right panel (the parameter settings of the pollutant cloud (Figure A1) or the data for reading particles of a pollutant cloud (Figure A2)) becomes enabled.

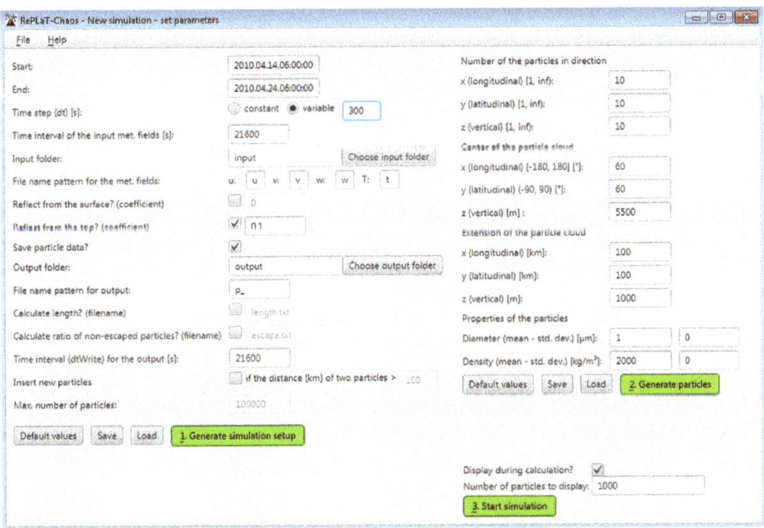

Figure A1. New simulation—set parameters. Starting a new simulation by setting the parameters of the simulation and the pollutant cloud.

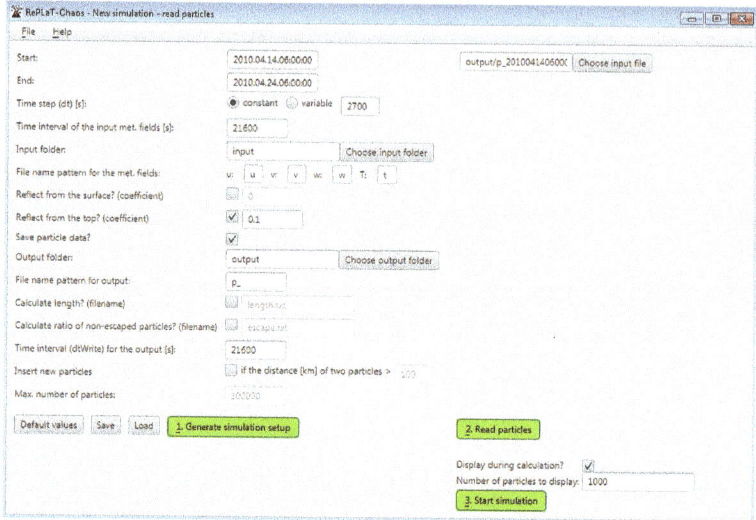

Figure A2. New simulation - read particles. Starting a new simulation by setting the parameters of the simulation and reading the particles of the pollutant clouds from file.

Appendix A.2.2. Setting the Parameters of the Pollutant Cloud

In case of choosing the New simulation—set parameters option (Figure A1) the user should set the following parameters of the particles which will fill a rectangular cuboid:

- Number of particles in direction x/y/z: the number of the particles in zonal, meridional and vertical direction (format: integers, greater or equal than 1).
- Center of the particle cloud x/y/z: zonal [°], meridional [°] and vertical [m] coordinate of the center of the pollutant cloud (format: real numbers in the given intervals)
- Extension of the particle cloud x/y/z: length of the sides of the rectangular cuboid representing the pollutant cloud in zonal [km], meridional [km] and vertical [m] direction (format: non-negative real)
- Diameter (mean–std.dev.): the diameter of the particles is log-normally distributed with the user given mean and standard deviation [μm] (format: non-negative real). As mentioned in Section 3.1, the diameter 0 μm corresponds to gas particles which are advected by the instantaneous velocity of the atmospheric flow at each time instant (their terminal velocity in Equation (6) is 0).
- Density (mean–std.dev.): the density of the particles is log-normally distributed with the user given mean and standard deviation [kg m^{-1}] (format: non-negative real).

By default, the fields are filled with the default setting parameters loaded from default/ default_particle_parameter_setup.txt. The usage of the Default, Save and Load buttons are analogous to the ones in the simulation setup panel.

For starting a simulation the generation of the pollutant cloud is required: the user can generate it by clicking the Generate particles button. If there is a wrong or empty text field, the problem is indicated by an alert window. Otherwise a pop-up window indicates that the pollutant cloud is generated (Number of particles: <particle number>.). Then the disabled bottom right panel (for setting the display properties of the simulation and starting the simulation calculation (Figure A1)) becomes enabled.

Appendix A.2.3. Reading the Particles of the Pollutant Cloud From File

In the case of reading the particles of the pollutant cloud (Figure A2), the user chooses the file containing the initial conditions of its particles by clicking the Choose input file button. The default file path is in the file default/default_particle_file_setup.txt. The formats and the values of data in the file should meet the requirements which are listed in Appendix A.1.3. The particle data are read from the file by clicking the Read particles button. In the case of wrong values an alert window indicates the problem: There were invalid data while reading from file <file> <wrong lines>. If every line is correct, a pop-up window indicates that the pollutant cloud is generated (Number of particles: <particle number>.). Then the disabled bottom right panel (for setting the display properties of the simulation and starting the simulation calculation (Figure A2)) becomes enabled.

This way of generating a pollutant cloud is worth applying when the user does not want to initialize the particles filling a rectangular cuboid (mentioned in the previous section) rather than the user needs particles with arbitrary positions. For example, in this way several different particle groups (i.e., different pollutant clouds) initialized at different locations can be tracked simultaneously.

Appendix A.2.4. Starting a New Simulation

In the bottom right panel (Figure A1 and Figure A2) the user should check whether if she/he wants to watch the spreading of the pollutant cloud during the simulation calculation (Display during calculation?) and if yes, how many particles of the pollutant cloud should be drawn (Number of particles to display, format: integer). By clicking the Start simulation button the simulation calculation starts and the progress (and the particle cloud if chosen) can be tracked in a new window (Figure A3).

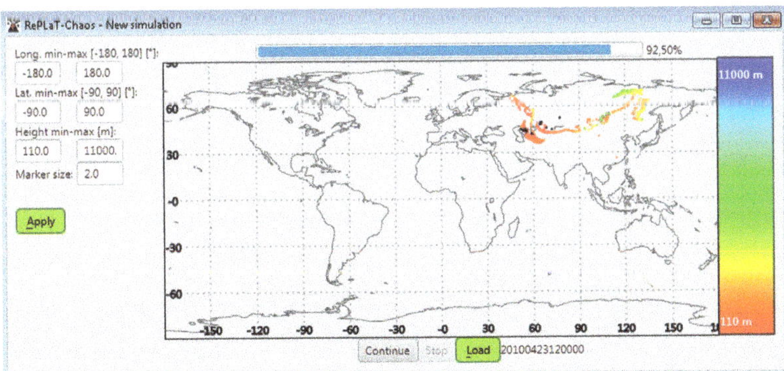

Figure A3. Example for tracking the calculation of the simulation with displaying the particle positions. The simulation and the pollutant cloud are initialized with the parameters in Figure A1. The colorbar indicates the altitude of the particles.

Appendix A.2.5. Setting the Display Properties of the Simulation

If the user has chosen the display option, the position of the particles of the pollutant cloud is displayed continuously colored according to the vertical coordinate of the particles on a map (Figure A3) during the simulation calculation. Otherwise only a progressbar is visible to show the percentage of the progress of the calculation and the corresponding date and time in the simulation. The user can stop and continue the calculation by clicking the Stop or Continue buttons, respectively, and she/he can load the particle data saved in files for replay by clicking the Load button. The longitudinal and latitudinal boundaries of the map, the vertical boundaries of the coloring and the marker size of the particles can be changed on the left side (formats: real numbers in the given intervals). The settings are applied by clicking the Apply button. If there is a wrong or empty text field, the problem is indicated by an alert window.

Appendix A.3. Replaying a Saved Simulation

A saved simulation can be loaded by clicking the Load simulation menu item in File menu on the new simulation screen or by clicking the Load button in the case of an ongoing simulation.

Then the parameters defining the saved simulation appear in a new window (Figure A4). For ongoing simulation text fields are filled with the parameters of the simulation, otherwise they are filled with the values loaded from the default/default_load_setup.txt. The following parameters should be given:

- Folder: folder of the files to be loaded. Clicking the Choose folder button it can be chosen or it can be written directly to the text field.
- File name pattern: the part of the name of the particle files before date and time.
- Length file: should length data be loaded, and if yes from which file. Clicking the Choose file button it can be chosen or it can be written directly to the text field.
- Escape file: should the data of the ratio of non-escaped particles be loaded, and if yes from which file. Clicking the Choose file button it can be chosen or it can be written directly to the text field. The format of the lines of the length and escape file should coincide with the requirements mentioned in Appendix A.1.3.
- Start: start of the display (format: <yyyy.MM.dd.hh:mm:ss>)
- End: end of the display (format: <yyyy.MM.dd.hh:mm:ss>)
- Time interval of the input files: time interval [s] between subsequent files to be displayed (format: integer)
- File for continents: the file which contains the coordinates ($[-\pi, \pi] \times [-\pi/2, \pi/2]$ [rad]) of the coastlines of the continents. Clicking the Choose file button it can be chosen, or it can be written directly to the text field.

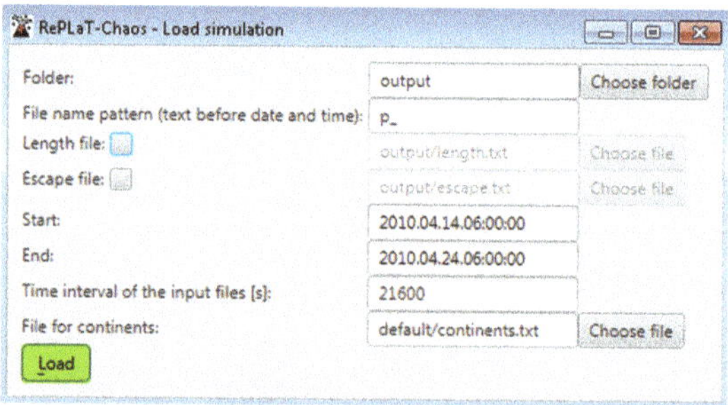

Figure A4. Setting simulation data for loading saved simulation.

The selected simulation is loaded by clicking the Load button. If the Length file/Escape file is checked, the time dependence of the natural logarithm of the length of the pollutant cloud/the time dependence of the natural logarithm of the ratio of the non-escaped particles also appears on the display panel (Figures A5 and A6). By default, the forward loop of the spreading of pollutant cloud is displayed according to the given Frame rate. The user can stop (Stop) and continue (Continue) the replay, and can move frame by frame the replay forward/backward by clicking the Previous/Next buttons. The instantaneous position of the pollutant cloud can be saved as an image by the Save image button. The properties of the display can be changed similarly as described in Appendix A.2.5 by clicking the Apply button. Beyond those options the speed of the animation (Frame rate) can be modified, too.

If the user has loaded length data/data for the ratio of the non-escaped particles from file, she/he can select a start and end date and time from two lists in the bottom of the panel in Figure A5 or Figure A6. By clicking the Calculate topological entropy/Calculate escape rate button a line is fitted to the graph between the given time instants using the least squares approach (Sections 3.2 and 3.3) and its slope (i.e., the topological entropy h (Figure A5))/$(-1)\times$ slope (i.e., the escape rate κ (Figure A6)) appears. The value of the obtained topological entropy h in Figure A5 means that in every $h^{-1} = 1.242$ days the length of the pollutant cloud stretches by a factor of e ≈ 2.718. Analogously, the value of the escape rate κ in Figure A6 implies that within $\kappa^{-1} = 3.745$ days after the start time of the fit only $e^{-1} \approx 0.368$ of the particles (which are non-escaped at the start time of the fit) are still in the air. The graphs can be saved by clicking the Save image button.

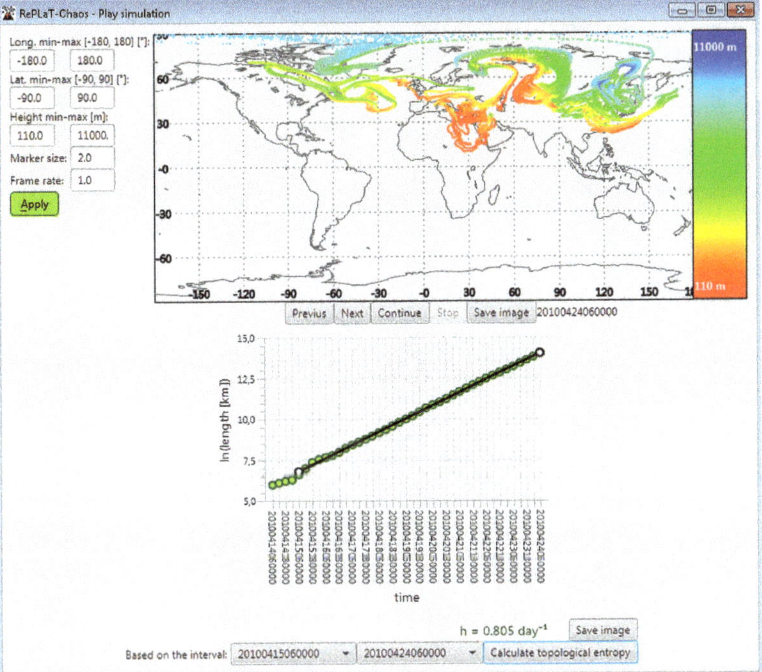

Figure A5. Displaying saved simulation and the time dependence of the length of a pollutant cloud. The simulation is initialized with the simulation parameters on the left of Figure A1 but with top and bottom reflection coefficients of 1 and inserting new particles if the distance of two particles is greater than 100 km. The pollutant cloud is initialized as a meridional line segment of 400 km at 47° N, 19° E and at the altitude of 5500 m. It consists of 1000 particles at the beginning. The particle radius is 0 μm.

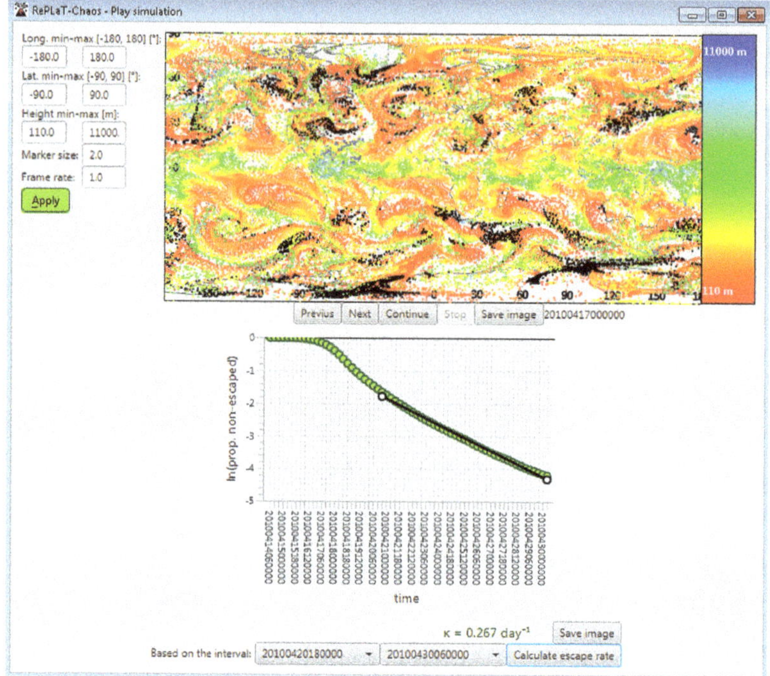

Figure A6. Displaying saved simulation and the time dependence of the ratio of non-escaped particles. The simulation is initialized with the simulation parameters on the left of Figure A1 but with an end date of "2010.04.30.06:00:00" and with calculating the ratio of non-escaped particles. The pollutant cloud is initialized as $250 \times 250 \times 1$ particles at 0° N, 0° E and at the altitude of 5500 m with an extension of 4×10^4 km $\times 2 \times 10^4$ km $\times 0$ m (i.e., covering the entire globe uniformly). The particle radius and density is 7 μm and 2000 kg m^{-3}, respectively.

References

1. Ulfarsson, G.F.; Unger, E.A. Impacts and Responses of Icelandic Aviation to the 2010 Eyjafjallajökull Volcanic Eruption: Case Study. *Transp. Res. Rec.* **2011**, *2214*, 144–151. [CrossRef]
2. Wilkinson, S.M.; Dunn, S.; Ma, S. The vulnerability of the European air traffic network to spatial hazards. *Nat. Hazards* **2012**, *60*, 1027–1036. [CrossRef]
3. Arason, P.; Petersen, G.; Bjornsson, H. Observations of the altitude of the volcanic plume during the eruption of Eyjafjallajökull, April–May 2010. *Earth Syst. Sci. Data* **2011**, *3*, 9–17. [CrossRef]
4. Stohl, A.; Prata, A.; Eckhardt, S.; Clarisse, L.; Durant, A.; Henne, S.; Kristiansen, N.I.; Minikin, A.; Schumann, U.; Seibert, P.; et al. Determination of time-and height-resolved volcanic ash emissions and their use for quantitative ash dispersion modeling: The 2010 Eyjafjallajökull eruption. *Atmos. Chem. Phys.* **2011**, *11*, 4333–4351. [CrossRef]
5. Schumann, U.; Weinzierl, B.; Reitebuch, O.; Schlager, H.; Minikin, A.; Forster, C.; Baumann, R.; Sailer, T.; Graf, K.; Mannstein, H.; et al. Airborne observations of the Eyjafjalla volcano ash cloud over Europe during air space closure in April and May 2010. *Atmos. Chem. Phys.* **2011**, *11*, 2245–2279. doi:10.5194/acp-11-2245-2011. [CrossRef]
6. Sicard, M.; Guerrero-Rascado, J.L.; Navas-Guzmán, F.; Preißler, J.; Molero, F.; Tomás, S.; Bravo-Aranda, J.A.; Comerón, A.; Rocadenbosch, F.; Wagner, F.; et al. Monitoring of the Eyjafjallajökull volcanic aerosol plume over the Iberian Peninsula by means of four EARLINET lidar stations. *Atmos. Chem. Phys.* **2012**, *12*, 3115–3130. [CrossRef]

7. Pappalardo, G.; Mona, L.; D'Amico, G.; Wandinger, U.; Adam, M.; Amodeo, A.; Ansmann, A.; Apituley, A.; Alados Arboledas, L.; Balis, D.; et al. Four-dimensional distribution of the 2010 Eyjafjallajökull volcanic cloud over Europe observed by EARLINET. *Atmos. Chem. Phys.* **2013**, *13*, 4429–4450. [CrossRef]
8. Burlakov, V.; Dolgii, S.; Nevzorov, A.; Samokhvalov, I.; Nasonov, S.; Zhivotenyuk, I.; El'nikov, A.; Nazarov, E.; Plusnin, I.; Shikhantsov, A. Traces of eruption of Eyjafjallajökull volcano according to data of lidar observations in Tomsk and Surgut. *Atmos. Ocean. Opt.* **2012**, *25*, 110–117. [CrossRef]
9. Zuev, V.V.; Burlakov, V.D.; Nevzorov, A.V.; Pravdin, V.L.; Savelieva, E.S.; Gerasimov, V.V. 30-year lidar observations of the stratospheric aerosol layer state over Tomsk (Western Siberia, Russia). *Atmos. Chem. Phys.* **2017**, *17*, 3067–3081. [CrossRef]
10. Petersen, G.; Bjornsson, H.; Arason, P.; Löwis, S.v. Two weather radar time series of the altitude of the volcanic plume during the May 2011 eruption of Grímsvötn, Iceland. *Earth Syst. Sci. Data* **2012**, *4*, 121–127. [CrossRef]
11. Tesche, M.; Glantz, P.; Johansson, C.; Norman, M.; Hiebsch, A.; Ansmann, A.; Althausen, D.; Engelmann, R.; Seifert, P. Volcanic ash over Scandinavia originating from the Grímsvötn eruptions in May 2011. *J. Geophys. Res. Atmos.* **2012**, *117*. [CrossRef]
12. Wilkins, K.; Western, L.; Watson, I. Simulating atmospheric transport of the 2011 Grímsvötn ash cloud using a data insertion update scheme. *Atmos. Environ.* **2016**, *141*, 48–59. [CrossRef]
13. Bowyer, T.W.; Biegalski, S.R.; Cooper, M.; Eslinger, P.W.; Haas, D.; Hayes, J.C.; Miley, H.S.; Strom, D.J.; Woods, V. Elevated radioxenon detected remotely following the Fukushima nuclear accident. *J. Environ. Radioact.* **2011**, *102*, 681–687. [CrossRef] [PubMed]
14. Leon, J.D.; Jaffe, D.; Kaspar, J.; Knecht, A.; Miller, M.; Robertson, R.; Schubert, A. Arrival time and magnitude of airborne fission products from the Fukushima, Japan, reactor incident as measured in Seattle, WA, USA. *J. Environ. Radioact.* **2011**, *102*, 1032–1038. [CrossRef] [PubMed]
15. MacMullin, S.; Giovanetti, G.; Green, M.; Henning, R.; Holmes, R.; Vorren, K.; Wilkerson, J. Measurement of airborne fission products in Chapel Hill, NC, USA from the Fukushima Dai-ichi reactor accident. *J. Environ. Radioact.* **2012**, *112*, 165–170. [CrossRef] [PubMed]
16. Masson, O.; Baeza, A.; Bieringer, J.; Brudecki, K.; Bucci, S.; Cappai, M.; Carvalho, F.; Connan, O.; Cosma, C.; Dalheimer, A.; et al. Tracking of airborne radionuclides from the damaged Fukushima Dai-ichi nuclear reactors by European networks. *Environ. Sci. Technol.* **2011**, *45*, 7670–7677. [CrossRef]
17. Bossew, P.; Kirchner, G.; De Cort, M.; De Vries, G.; Nishev, A.; De Felice, L. Radioactivity from Fukushima Dai-ichi in air over Europe; part 1: Spatio-temporal analysis. *J. Environ. Radioact.* **2012**, *114*, 22–34. [CrossRef]
18. Stohl, A.; Seibert, P.; Wotawa, G.; Arnold, D.; Burkhart, J.; Eckhardt, S.; Tapia, C.; Vargas, A.; Yasunari, T. Xenon-133 and caesium-137 releases into the atmosphere from the Fukushima Dai-ichi nuclear power plant: Determination of the source term, atmospheric dispersion, and deposition. *Atmos. Chem. Phys.* **2012**, *12*, 2313–2343. [CrossRef]
19. Manolopoulou, M.; Vagena, E.; Stoulos, S.; Ioannidou, A.; Papastefanou, C. Radioiodine and radiocesium in Thessaloniki, Northern Greece due to the Fukushima nuclear accident. *J. Environ. Radioact.* **2011**, *102*, 796–797. [CrossRef]
20. Pittauerová, D.; Hettwig, B.; Fischer, H.W. Fukushima fallout in Northwest German environmental media. *J. Environ. Radioact.* **2011**, *102*, 877–880. [CrossRef]
21. Bikit, I.; Mrda, D.; Todorovic, N.; Nikolov, J.; Krmar, M.; Veskovic, M.; Slivka, J.; Hansman, J.; Forkapic, S.; Jovancevic, N. Airborne radioiodine in northern Serbia from Fukushima. *J. Environ. Radioact.* **2012**, *114*, 89–93. [CrossRef] [PubMed]
22. Castruccio, A.; Clavero, J.; Segura, A.; Samaniego, P.; Roche, O.; Le Pennec, J.L.; Droguett, B. Eruptive parameters and dynamics of the April 2015 sub-Plinian eruptions of Calbuco volcano (southern Chile). *Bull. Volcanol.* **2016**, *78*, 62. [CrossRef]
23. Romero, J.; Morgavi, D.; Arzilli, F.; Daga, R.; Caselli, A.; Reckziegel, F.; Viramonte, J.; Díaz-Alvarado, J.; Polacci, M.; Burton, M.; et al. Eruption dynamics of the 22–23 April 2015 Calbuco Volcano (Southern Chile): Analyses of tephra fall deposits. *J. Volcanol. Geotherm. Res.* **2016**, *317*, 15–29. [CrossRef]
24. Reckziegel, F.; Bustos, E.; Mingari, L.; Báez, W.; Villarosa, G.; Folch, A.; Collini, E.; Viramonte, J.; Romero, J.; Osores, S. Forecasting volcanic ash dispersal and coeval resuspension during the April–May 2015 Calbuco eruption. *J. Volcanol. Geotherm. Res.* **2016**, *321*, 44–57. [CrossRef]

25. Ivy, D.J.; Solomon, S.; Kinnison, D.; Mills, M.J.; Schmidt, A.; Neely, R.R. The influence of the Calbuco eruption on the 2015 Antarctic ozone hole in a fully coupled chemistry-climate model. *Geophys. Res. Lett.* **2017**, *44*, 2556–2561. [CrossRef]

26. Corsaro, R.A.; Andronico, D.; Behncke, B.; Branca, S.; Caltabiano, T.; Ciancitto, F.; Cristaldi, A.; De Beni, E.; La Spina, A.; Lodato, L.; et al. Monitoring the December 2015 summit eruptions of Mt. Etna (Italy): Implications on eruptive dynamics. *J. Volcanol. Geotherm. Res.* **2017**, *341*, 53–69. [CrossRef]

27. Pompilio, M.; Bertagnini, A.; Del Carlo, P.; Di Roberto, A. Magma dynamics within a basaltic conduit revealed by textural and compositional features of erupted ash: The December 2015 Mt. Etna paroxysms. *Sci. Rep.* **2017**, *7*, 4805. [CrossRef]

28. Athanassiadou, M. The Mt Etna SO2 eruption in December 2015—The view from space. *Weather* **2016**, *71*, 273–279. [CrossRef]

29. Haszpra, T.; Tél, T. Escape rate: A Lagrangian measure of particle deposition from the atmosphere. *Nonlinear Process. Geophys.* **2013**, *20*, 867–881. [CrossRef]

30. Haszpra, T.; Herein, M. Ensemble-based analysis of the pollutant spreading intensity induced by climate change. *Sci. Rep.* **2019**, *9*, 3896. [CrossRef]

31. Haszpra, T.; Tél, T. Topological entropy: A Lagrangian measure of the state of the free atmosphere. *J. Atmos. Sci.* **2013**, *70*, 4030–4040. [CrossRef]

32. Dutton, E.G.; Christy, J.R. Solar radiative forcing at selected locations and evidence for global lower tropospheric cooling following the eruptions of El Chichon and Pinatubo. *Geophys. Res. Lett.* **1992**, *19*, 2313–2316. [CrossRef]

33. Kerr, R.A. Pinatubo global cooling on target. *Science* **1993**, *259*, 594–595.

34. Self, S.; Zhao, J.X.; Holasek, R.E.; Torres, R.C.; King, A.J. The atmospheric impact of the 1991 Mount Pinatubo eruption. In *Fire and Mud: The Eruptions and Lahars of Mount Pinatubo, Philippines*; C.G. Newhall, R.P., Ed.; University of Washington Press: Seattle, WA, USA, 1996; pp. 1089–1116.

35. Ott, E. *Chaos in Dynamical Systems*; Cambridge University Press: New York, NY, USA, 1993; p. 385.

36. Tél, T.; Gruiz, M. *Chaotic Dynamics: An Introduction Based on Classical Mechanics*; Cambridge University Press: Cambridge, UK, 2006; p. 393.

37. Lai, Y.C.; Tél, T. *Transient Chaos: Complex Dynamics on Finite Time Scales*; Springer-Verlag New York: New York, NY, USA, 2011; Volume 173.

38. Tél, T. The joy of transient chaos. *Chaos Interdiscip. J. Nonlinear Sci.* **2015**, *25*, 097619. [CrossRef]

39. Haszpra, T. Intensification of Large-Scale Stretching of Atmospheric Pollutant Clouds due to Climate Change. *J. Atmos. Sci.* **2017**, *74*, 4229–4240. [CrossRef]

40. Stein, A.; Draxler, R.R.; Rolph, G.D.; Stunder, B.J.; Cohen, M.; Ngan, F. NOAA's HYSPLIT atmospheric transport and dispersion modeling system. *Bull. Am. Meteorol. Soc.* **2015**, *96*, 2059–2077. [CrossRef]

41. Rolph, G.; Stein, A.; Stunder, B. Real-time environmental applications and display sYstem: READY. *Environ. Model. Softw.* **2017**, *95*, 210–228. [CrossRef]

42. Miltenberger, A.K.; Pfahl, S.; Wernli, H. An online trajectory module (version 1.0) for the nonhydrostatic numerical weather prediction model COSMO. *Geosci. Model Dev.* **2013**, *6*, 1989–2004. doi:10.5194/gmd-6-1989-2013. [CrossRef]

43. Sprenger, M.; Wernli, H. The LAGRANTO Lagrangian analysis tool—Version 2.0. *Geosci. Model Dev.* **2015**, *8*, 2569–2586. doi:10.5194/gmd-8-2569-2015. [CrossRef]

44. Pisso, I.; Sollum, E.; Grythe, H.; Kristiansen, N.; Cassiani, M.; Eckhardt, S.; Arnold, D.; Morton, D.; Thompson, R.L.; Groot Zwaaftink, C.D.; et al. The Lagrangian particle dispersion model FLEXPART version 10.3. *Geosci. Model Dev. Discuss.* **2019**, *2019*, 1–67. doi:10.5194/gmd-2018-333. [CrossRef]

45. Costa, A.; Macedonio, G.; Folch, A. A three-dimensional Eulerian model for transport and deposition of volcanic ashes. *Earth Planet. Sci. Lett.* **2006**, *241*, 634–647. [CrossRef]

46. Folch, A.; Costa, A.; Macedonio, G. FALL3D: A computational model for volcanic ash transport and deposition. *Comput. Geosci.* **2009**, *35*, 1334–1342. [CrossRef]

47. Folch, A.; Costa, A.; Macedonio, G. FPLUME-1.0: An integral volcanic plume model accounting for ash aggregation. *Geosci. Model Dev.* **2016**, *9*, 431–450. [CrossRef]

48. RePLaT-Chaos. Available online: http://theorphys.elte.hu/fiztan/volcano/#full (accessed on 21 December 2019).

49. Chandler, M.A.; Richards, S.J.; Shopsin, M. EdGCM: Enhancing climate science education through climate modeling research projects. In Proceedings of the 85th Annual Meeting of the American Meteorological Society, 14th Symposium on Education, San Diego, CA, USA, 8–14 January 2005; p. P1.

50. Fraedrich, K.; Jansen, H.; Kirk, E.; Luksch, U.; Lunkeit, F. The Planet Simulator: Towards a user friendly model. *Meteorol. Z.* **2005**, *14*, 299–304. [CrossRef]

51. Newhouse, S.; Pignataro, T. On the estimation of topological entropy. *J. Stat. Phys.* **1993**, *72*, 1331–1351. [CrossRef]

52. Thiffeault, J.L. Braids of entangled particle trajectories. *Chaos* **2010**, *20*, 017516–017516. [CrossRef]

53. Budišić, M.; Thiffeault, J.L. Finite-time braiding exponents. *Chaos* **2015**, *25*, 087407. [CrossRef]

54. Haszpra, T.; Horányi, A. Some aspects of the impact of meteorological forecast uncertainties on environmental dispersion prediction. *Idojaras* **2014**, *118*, 335–347.

55. Haszpra, T. Intricate features in the lifetime and deposition of atmospheric aerosol particles. *Chaos Interdiscip. J. Nonlinear Sci.* **2019**, *29*, 071103. [CrossRef]

56. Haszpra, T.; Tél, T. Volcanic ash in the free atmosphere: A dynamical systems approach. In *Journal of Physics: Conference Series*; IOP Publishing: Bristol, UK, 2011; Volume 333, p. 012008.

57. Haszpra, T.; Tél, T. Individual particle based description of atmospheric dispersion: A dynamical systems approach. In *The Fluid Dynamics of Climate*; Springer-Verlag Wien: Wien, Austria, 2016; pp. 95–119.

58. Sutherland, W. LII. The viscosity of gases and molecular force. *Philos. Mag. Ser.* **1893**, *36*, 507–531. doi:10.1080/14786449308620508. [CrossRef]

59. Stohl, A.; Wotawa, G.; Seibert, P.; Kromp-Kolb, H. Interpolation errors in wind fields as a function of spatial and temporal resolution and their impact on different types of kinematic trajectories. *J. Appl. Meteorol.* **1995**, *34*, 2149–2165. [CrossRef]

60. Flentje, H.; Claude, H.; Elste, T.; Gilge, S.; Köhler, U.; Plass-Dülmer, C.; Steinbrecht, W.; Thomas, W.; Werner, A.; Fricke, W. The Eyjafjallajökull eruption in April 2010–detection of volcanic plume using in-situ measurements, ozone sondes and lidar-ceilometer profiles. *Atmos. Chem. Phys.* **2010**, *10*, 10085–10092. [CrossRef]

61. Belosi, F.; Santachiara, G.; Prodi, F. Eyjafjallajökull volcanic eruption: Ice nuclei and particle characterization. *Atmos. Clim. Sci.* **2011**, *1*, 48.

62. Campanelli, M.; Estelles, V.; Smyth, T.; Tomasi, C.; Martìnez-Lozano, M.; Claxton, B.; Muller, P.; Pappalardo, G.; Pietruczuk, A.; Shanklin, J.; et al. Monitoring of Eyjafjallajökull volcanic aerosol by the new European Skynet Radiometers (ESR) network. *Atmos. Environ.* **2012**, *48*, 33–45. [CrossRef]

63. Revuelta, M.; Sastre, M.; Fernández, A.; Martín, L.; García, R.; Gómez-Moreno, F.; Artíñano, B.; Pujadas, M.; Molero, F. Characterization of the Eyjafjallajökull volcanic plume over the Iberian Peninsula by lidar remote sensing and ground-level data collection. *Atmos. Environ.* **2012**, *48*, 46–55. [CrossRef]

64. RePLaT-Chaos-edu. Available online: http://theorphys.elte.hu/fiztan/volcano/#edu (accessed on 21 December 2019).

65. Haszpra, T.; Kiss, M.; Izsa, É. RePLaT-Chaos-edu: An interactive educational tool for secondary school students for the illustration of the spreading of volcanic ash clouds. *J. Phys. Conf. Ser.* **2019**, submitted.

66. ECMWF Public Datasets. Available online: http://apps.ecmwf.int/datasets/ (accessed on 21 December 2019).

67. Holton, J.R. *An Introduction to Dynamic Meteorology*, 4th ed.; Academic Press: Cambridge, MA, USA, 2004; p. 535.

 *atmosphere*

Article

Capturing Plume Rise and Dispersion with a Coupled Large-Eddy Simulation: Case Study of a Prescribed Burn

Nadya Moisseeva * and Roland Stull

Department of Earth, Ocean and Atmospheric Sciences, The University of British Columbia, Vancouver, BC V6T 1Z4, Canada; rstull@eoas.ubc.ca
* Correspondence: nmoisseeva@eoas.ubc.ca

Received: 16 August 2019; Accepted: 23 September 2019; Published: 25 September 2019

Abstract: Current understanding of the buoyant rise and subsequent dispersion of smoke due to wildfires has been limited by the complexity of interactions between fire behavior and atmospheric conditions, as well as the uncertainty in model evaluation data. To assess the feasibility of using numerical models to address this knowledge gap, we designed a large-eddy simulation of a real-life prescribed burn using a coupled semi-emperical fire–atmosphere model. We used observational data to evaluate the simulated smoke plume, as well as to identify sources of model biases. The results suggest that the rise and dispersion of fire emissions are reasonably captured by the model, subject to accurate surface thermal forcing and relatively steady atmospheric conditions. Overall, encouraging model performance and the high level of detail offered by simulated data may help inform future smoke plume modeling work, plume-rise parameterizations and field experiment designs.

Keywords: wildfire plume rise; smoke modeling; large eddy simulation; emissions dispersion; WRF-SFIRE; RxCADRE

1. Introduction

Wildland fires cover a broad range of spatiotemporal scales and are shaped by the complex interaction of fuel, terrain, and meteorological conditions. While scientific understanding of wildland fires and associated smoke plumes are central to the successful mitigation of negative air-quality impacts, the complex and highly dynamic nature of fires presents a challenge for modeling. Existing smoke plume prediction models span a vast range of complexity from simple empirical relations to the more recent coupled fire–atmosphere numerical approaches. Often the choice of model is dictated by the context of its application, subject to the trade-off between fidelity and timely execution.

Large eddy simulation (LES) is a method that uses computational fluid dynamics at a very fine spatial and temporal resolution to simulate a wide range of scales of atmospheric motions down to the size of large turbulent eddies. The Weather Research and Forecasting Model, combined with a semi-empirical fire-spread algorithm (WRF-SFIRE), allows two-way coupling between an LES and a fire behavior model [1–3]. Several studies have examined the ability of WRF-SFIRE to capture the ground-spread behavior of a fire line, near-surface temperatures and winds [1,4,5]. Large-scale simulations of two real fires were carried out by Kochanski et al. [6], comparing modeled plume tops with satellite data. To the authors' best knowledge, very limited consideration has been given to assessing the ability of WRF-SFIRE to simulate wildfire smoke plume dynamics, and vertical rise and distribution of emissions on a local scale. This is the central motivation for this study.

As noted by Mallia et al. [7], there is a general lack of research focusing on modeling the vertical distribution of smoke emissions as a result of wildfires. This knowledge gap can, in part, be explained by the difficulty of constraining potential sources of error in both inputs and models themselves.

Until recently, evaluation of coupled fire–atmosphere models required a combination of studies, as no dataset was complete enough to rigorously constrain the problem [1]. Comprehensive field observations were needed to better our understanding of the interactions between fuels, fire behavior and meteorology.

In response, the Prescribed Fire Combustion and Atmospheric Dynamics Research Experiment (RxCADRE) was designed to address this critical research need [8]. The project brought together researchers from a wide range of disciplines to collect data on fuel, meteorology, fire behavior, energy, smoke emissions and fire effects. Simultaneous measurement of multiple fire aspects on the same prescribed burns provided a detailed model evaluation dataset, while also capturing the effects of fire–atmosphere coupling [9]. Data from this comprehensive experiment offers a unique opportunity to assess the accuracy WRF-SFIRE simulated plume rise and dynamics.

In addition, the modeling work we present may help inform future observational studies by identifying key aspects of experimental design. Note that the focus of this work is the evaluation of the LES ability to capture the atmospheric response to a simulated fire of known bulk properties, rather than the fire behavior itself. Effectively, the work aims to validate the relationship between the simulated surface forcing due to a fire and the resultant turbulent convection.

The findings are likely to be of interest for atmospheric and air-quality modelers, as detailed measurements of wildfire smoke plumes are scarce. "Synthetic" plume data from an LES would provide researchers with an alternative resource for validating their models. Therefore, the broad goal of this work is to assess the utility of WRF-SFIRE for improving plume rise and dispersion parameterizations.

2. Methods

2.1. Observational Data

The RxCADRE campaign consisted of 10 operational and 6 small replicate prescribed fires. Collected data are accessible via a US Forest Service online repository, as referenced below. Smoke dispersion and emissions measurements are available for three large fires: L1G and L2G grass fires and L2F sub-forest canopy surface fire. For the purpose of model evaluation, we selected L2G (10 November 2012) for our case study, based on its reported uniformity and consistency of flame propagation [10]. Figure 1 shows a sample snapshot of the burn plot during the ignition. The overall meteorological conditions and instrumental design of the L2G experimental burn are described in detail in [9]. The individual datasets obtained from the US Forest Service online archive used for this study are summarized below.

Georeferencing data, including plot location and burn perimeters, are available from Hudak and Bright [11]. Analysis of fire rate of spread (ROS) and intensity as well as a detailed description of three Highly Instrumented Plots (HIPs) used to produce the estimates can be found in [10]. Locations of HIPs are available from Hudak et al. [12]. HIP1, used for this evaluation, is shown in Figure 1. Near-surface wind and temperature sonic anemometer time series for in-situ and background locations are available from Seto and Clements [13,14]. Ignitions timing and locations were obtained from field-grade GPS units, mounted on-board firing vehicles [15]. Fuel data used for this evaluation study included photographs of pre-burn samples, as well as measurements of size, loading and moisture content of species groups. Data collection methodology is detailed in [16]. Dispersion and emissions measurements included volume-mixing ratio of CO_2, CO, CH_4, and water vapor at a rate of 2 s, obtained from aircraft-mounted sensors [17]. The georeferenced data consisted of horizontal transects at multiple elevations, as well as "corkscrew" and "parking garage" flight profiles.

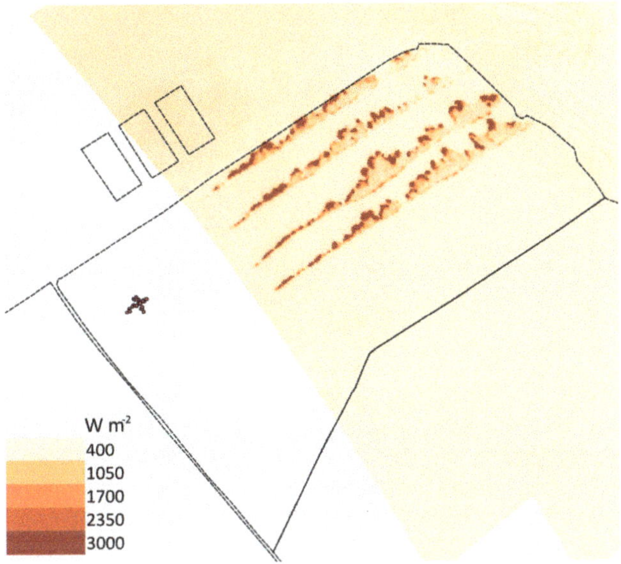

Figure 1. Long wave infra-red (LWIR) image of L2G lot during ignition (12:32:02 CST) with dashed black lines denoting burn perimeters. Red scatter points correspond to Highly Instrumented Plot (HIP) #1 fire behavior packages (FBP), each containing a system of airflow, temperature and energy sensors.

2.2. Numerical Setup

WRF-SFIRE [3] was configured in idealized LES mode. One of the primary advantages of using this model is that it allows for two-way coupling between the fire and the atmosphere. While WRF-SFIRE does not model combustion directly, the spread and intensity of the fire are parameterized using a semi-empirical approach. The latent heat flux is computed based on the fuel consumption and stoichiometric combustion of cellulose. Heat and moisture fluxes from the simulated burn provide forcing to the atmosphere, which in turn influences fire behavior.

A 10.4 km × 14 km domain with 40 m horizontal grid spacing, 3000 m model top and 51 hyperbolically stretched vertical levels was initialized using the 10:00 CST (16:00 UTC) sounding [9]. While this may appear to be a shallow domain compared to mesoscale ("Real") WRF simulations, the choice is substantially higher than that found in existing published WRF-SFIRE evaluations [1,4,18]. Five lowest model grid centers were located at approximately 8 m, 24 m , 42 m, 60 m and 80 m above ground level (AGL). The simulation was allowed to spin up for 2 h 23 min prior to ignition at ~12:23 CST (time varied slightly for different fire lines). To aid the formation of buoyancy-driven ambient background turbulence typical for a daytime boundary layer (BL), a lower-boundary surface thermal flux (tke_heat_flux) was imposed. The value was estimated from the sonic anemometer time series of vertical wind velocity and temperature over the time period leading up to ignition. As shown in Figure 2, based on the measurements, the ambient background surface heat flux remained fairly constant over the entire spin-up period. Hence, the lower-boundary surface forcing was idealized for the LES simulation as being uniform in space and constant in time. We used full surface initialization (sfc_full_init =.true.), with the lower boundary moisture flux and surface roughness characteristics set to standard USGS values for "Grassland" land use category.

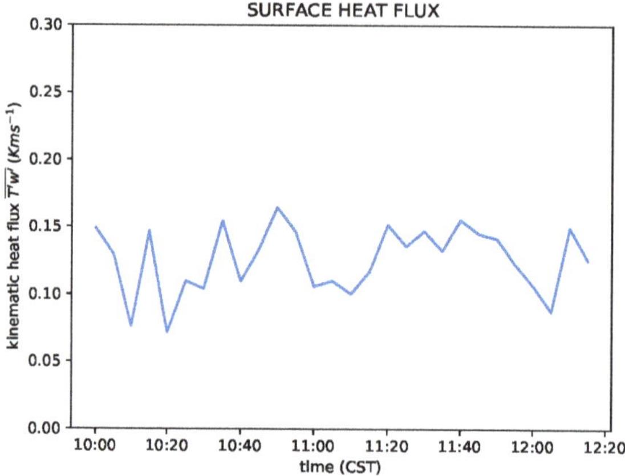

Figure 2. Five-minute averaged kinematic surface heat flux $\overline{T'w'}$ derived from 1 Hz wind and temperature sonic anemometer time series of the background ambient environment.

To help trigger convection in a horizontally uniform initial domain a small temperature perturbation "bubble" was added (see namelist.input_spinup in the Supplementary Materials). With periodic boundary conditions, near-stationary turbulence spectrum was achieved within ~40 min of run start. The well-mixed modeled BL continued to turn over and warm for a total of 2 h 23 min (10:00:00 CST–12:23:00 CST). Restart file generated at 12:23:00 CST was used as initial conditions for the main burn simulation (12:23:00 CST–13:12:00 CST), ensuring the fire was ignited into a well developed BL. Other key configuration details can be found in Table 1, as well as in the complete namelist initialization files provided as Supplementary Materials.

Table 1. Key parameters of numerical domain setup.

Simulation Parameter	Value/Description
Model version	24 May 2019 (git https://github.com/openwfm/wrf-fire/tree/ced5955b23cfa9bc0f937783c1c63ff7aa1bc2fa)
Horizontal grid spacing	40 m
Domain size	260 grids (east-west) × 350 grids (north-south)
Time step	0.1 s
Model top	3000 m AGL
Spinup timing	10:00:00–12:23:00 CST (CST = UTC − 6 h)
Fire (restart) simulation timing	12:23:00–13:12:00 CST
Sub-grid scale closure	1.5 TKE (TKE = Turbulence kinetic energy)
Lateral boundary conditions	periodic
Surface physics	Monin–Obukhov similarity (sf_sfclay_physics = 1)
Land surface model	thermal diffusion (sf_surface_physics = 1)
Surface heat flux	160 W m^{-2} (tke_heat_flux = 0.13)

Following the LES spin up, the northwestern half of the simulated L2G lot was ignited with four roughly parallel fire lines mimicking strip head fire method used during the real-life burn (Figure 1). During the campaign, the prescribed burn was ignited with drip torches attached to moving all-terrain vehicles (ATVs). Using GPS data from these vehicles (available from [15]), we extracted the locations of start and end points of the four fire lines, as well as their individual start and end ignition times. While the real-life ignition process was not perfectly uniform in time, the modeled fire lines were approximated as being ignited at a constant speed, such that the time and location of the start and end points matched those of the real burn (see Animation S1 in the Supplementary Materials). Timing varied slightly for each of the four modeled fire lines (see namelist.input_main in the Supplementary Materials). We approximated the ignitions as straight lines between observed start and end points, as the ATVs' deflections from a straight path during the real burn remained within a single atmospheric grid in our modeled domain.

Ignited cells in WRF-SFIRE proceeded to spread, while each fire line continued to advance until reaching the opposite end of the L2G lot. Subsequent upwind ignitions of the remaining lot area were excluded to reduce the computational load of the simulation. Taking into account the downwind location and timing of smoke plume observations, this simplification should have no effect on the proposed evaluation. The simulation was allowed to proceed for 49 min, until the emissions reached the downwind end of the domain.

Summary of fire and fuel parameters can be found in Table 2. Based on photographs and average measurements of fuel size, composition and type, we determined Anderson's fuel Category 1 (short grass) [19] to be the best fit for L2G ground cover. Actual burn perimeters were used to mask the remaining domain as containing no fuel to prevent spread of the simulated burn outside of the burn lot. We replaced the standard fuel loading and depth associated with Type 1 fuels with average measured values of 0.267 kg m^{-2} and 0.18 m, respectively. Surface dead fuel moisture content was set to 8.46% based on observations. Heat of combustion of dry fuel was adjusted to 1.64×10^7 J kg^{-1} as per estimates for grasslands provided by Overholt et al. [20].

Table 2. Details of fire and ignition parameters in LES setup.

Simulation Parameter	Value
Fire mesh refinement	10
Ignition duration	12:23–12:36 CST (varied for each fire line)
Rate of spread during ignition	0.2 m s^{-1}
Fuel category	1 (short grass)
Surface dead fuel moisture	8.46%
Heat of combustion of dry fuel	1.64×10^7 J kg^{-1}

As the central goal of this work is to evaluate the model's ability to capture wildfire smoke plume dynamics, we did not incorporate chemistry coupling into the simulation. Modeled "smoke plume" was represented by two passive tracers released proportionally to the mass and type of fuel burned. The rate of release for each tracer representing CO and CO_2 was controlled by assigned emission factors, based on values for grasslands provided by [21] (see namelist.fire_emissions in the Supplementary Materials).

3. Results

The overall evolution of the simulated L2G burn and the associated smoke plume is best visualized with a 3D animation (see Animation S1 in the Supplementary Materials). The Supplementary Materials also includes an animated view of the cross-wind modeled CO_2 mixing ratio (Animation S2). The latter demonstrates the ability of the LES to capture common plume behavior. As seen in the animation, the initial rise of moist buoyant air results in a temporary overshoot of the equilibrium plume height, followed by the gradual settling of the plume to its final injection height near the top of the boundary layer for this case. While the ability of WRF-SFIRE to qualitatively capture typical plume dynamics is reassuring, the following sections take a more quantitative approach to model evaluation.

3.1. Fire Behavior

Prior to evaluating the ability of WRF-SFIRE to capture plume rise and dispersion, it is important to ensure that the model is able to reasonably simulate fire behavior. Initial surface and fuel conditions have the potential to strongly impact fire growth and intensity, and, hence, affect the location and buoyancy of the smoke plume. As noted in Section 1, our approach does not constitute a comprehensive fire behavior evaluation study, but rather aims to ensure that WRF-SFIRE captures the bulk properties of combustion and supplies a reasonable surface forcing to the simulated atmosphere.

Our evaluation is based on the analysis of fire energy transport of RxCADRE observational data for L2G burn carried out by Butler et al. [10]. The study provides measurement-based values as well as error margins for ROS, and peak and average heat fluxes of the fire, which we use to assess the performance of the semi-empirical fire algorithm driving our LES simulation. Figure 3a,b compares LES-derived average and peak total heat fluxes for HIP1 and entire burn area over the flaming period with observations. For HIP1 point-to-point comparison, we use output from the nearest modeled grid points. L2G average observed values include measurements from all three HIP lots. The corresponding simulated estimates are calculated using the entire burn area (roughly half of the L2G lot).

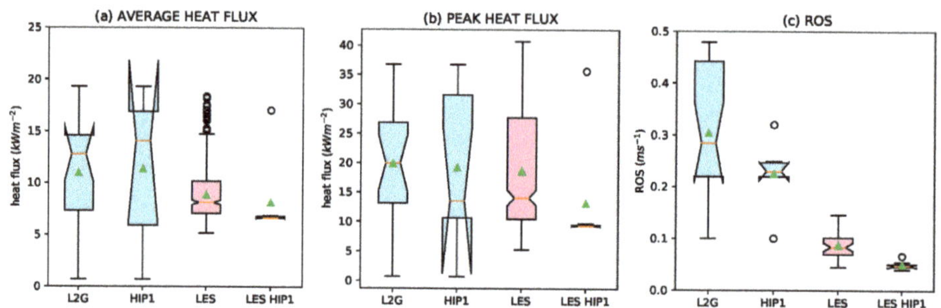

Figure 3. Comparison of observed (blue) and modeled (red) fire behavior. The box and whiskers span interquartile range (IQR) and $1.5 \times$ IQR, respectively, with the notch denoting the 95% confidence interval of the median (median $\pm 1.57 \times$ IQR/$n^{1/2}$). Red line and green triangle correspond to median and mean, respectively. (**a**) Average heat flux during flaming period. (**b**) Peak fire heat flux during flaming period. (**c**) Rate of spread.

The start and end times of the flaming period are defined as simulation frames at which total heat flux at the location exceeded 5 kW m^{-2} [10]. For both burn-wide and point comparisons, the flaming period is determined separately for each individual grid point. Only ignited grids are included in the analysis. This approach allows us to mimic the analysis performed by Butler et al. [10] in the absence of true combustion modeling in WRF-SFIRE.

For the entire burn area the observed mean and peak heat fluxes associated with the fire (not the background environment) are 11 kW m^{-2} and 20 kW m^{-2}, compared to LES-derived values of 8.9 kW m^{-2} and 19 kW m^{-2}, respectively. For HIP1 lot the corresponding values were 11.4 kW m^{-2} and 19.4 kW m^{-2} (observed) versus 8.2 kW m^{-2} and 13 kW m^{-2} (modeled). Note that, due to close proximity of the HIP1 sensors to each other, four out of seven of them fall into the same atmospheric grid within the modeled domain. Modeled HIP1 averages should therefore be treated with caution, as they consist of only four unique values. Moreover, the large spread of observed HIP1 heat fluxes renders the differences between model and measurements not statistically significant. Overall, the results shown in Figure 3 suggest that on average the surface thermal forcing to the modeled atmosphere due to the fire is reasonably captured by the model, subject to a slight negative bias (significant and non-significant for average and peak heat fluxes, respectively).

Observed rates of spread during the L2G burn were estimated using two methods in the study by Butler et al. [10]: flame arrival time from ignition and video images. The former approach takes into account the ignition time of the nearest fire line (perpendicular to fire advance vector) and the distance to the individual HIP1 sensors. The resultant values appear to have lower associated uncertainty than the latter image-derived method. To ensure consistency, we mimicked the above methodology in our simulated domain. Using the high-resolution fire domain, we calculated the upwind distance between each HIP1 point and the ignition line and the time it took the flame to reach each sensor location. To estimate ROS for the entire burn area, we created a mid-fire cross-section of 50 point-pairs between second and third ignition lines. Similar to the approach above, we derived the distance and flame travel time for each pair to calculate ROS. As shown in Figure 3c, mean LES-based HIP1 and L2G ROS values of 0.049 m s^{-1} and 0.087 m s^{-1} are significantly lower then the corresponding observed rates of spread (0.23 m s^{-1} and 0.30 m s^{-1}, respectively). Possible implications and sensitivity of our results to this deficiency are addressed in Section 4.

3.2. Plume Dynamics

Airborne emissions data collected during RxCADRE campaign is central to our evaluation of WRF-SFIRE's ability to capture plume rise and dispersion. The emissions dataset [17] contains smoke plume entry and exit points along the flight path, which were calculated using background CO baseline concentrations. The measurements were taken along horizontal transects passing through the plume at various vertical levels ("parking garage" profile), beginning close to the ground and moving towards the top of the plume, for a total of 9 crossings.

The identified in-plume segments were then compared with modeled CO mixing ratios along the same flight path extracted from the geo- and time-referenced LES domain. Figure 4 shows the time series of the flight path simulated emissions, overlaid with observations-derived plume segments. The results suggest good overall agreement in both location and timing between the modeled and observed emissions dispersion throughout majority of the BL depth. The coinciding model CO peaks and observed smoke segments indicate that the horizontal width of the smoke plume is well represented in the model. Potential shortcomings include excess smoke near the ground, as suggested by the early peaks (12:36 and 12:40 CST) not identified as a plume crossing, as well as a slight skew of the overall smoke distribution towards higher levels. A small phase shift appears in the modeled peaks toward the later parts of the simulation (12:50 CST and beyond).

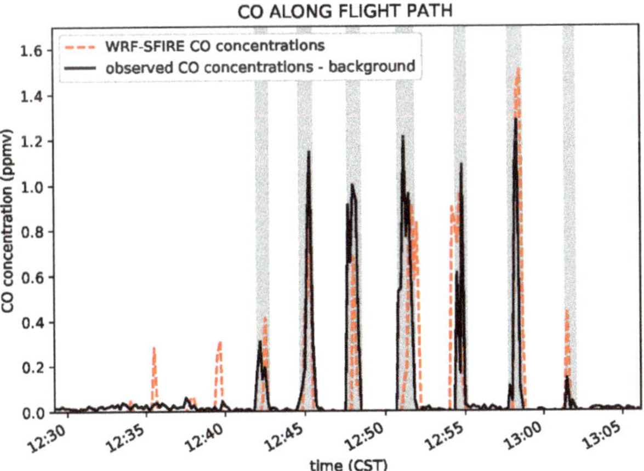

Figure 4. Simulated CO mixing ratio along RxCADRE flight path. Red dashed and solid black lines correspond to LES-derived and observed values, respectively. Gray shading indicates observed smoke time periods (not magnitudes) as identified from CO measurements along the flight path.

To evaluate the vertical distribution of WRF-SFIRE emissions, we compared the model-generated CO_2 concentrations with airborne measurements obtained during the "parking garage" and "corkscrew" (spiral ascent or descent) maneuvers. As shown in Figure 5a, there is a good overall agreement in injection heights for fire-generated emissions during the earlier "parking garage" profile. Plume top is accurately captured. Modeled concentrations tend to have a negative bias of ∼5 ppmv throughout the bulk of the plume thickness (500–1300 m), and be slightly over-predicted for the very top and bottom of the smoke column (at 400 m and 1500 m).

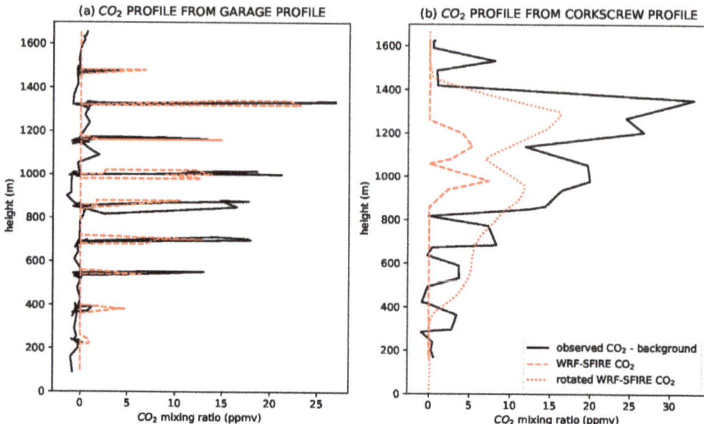

Figure 5. Observed (black) and modeled (red) vertical CO_2 emissions distribution during: (**a**) "parking garage" maneuver; and (**b**) corkscrew maneuver.

The "corkscrew" profile corresponds to a time near the very end of our simulation. As shown in Figure 5b, the band of modeled emissions appears to be very narrow and severely under-predicts the smoke concentrations. We discuss possible reasons for this behavior in Section 4.

4. Discussion

The aim of our WRF-SFIRE evaluation was to assess its ability to capture fire-generated emissions in the context of air quality. Hence, we examined the implications of the above results based on their potential applications for wildfire smoke plume rise and dispersion modeling. The following sections discuss model performance and accuracy from the perspective of atmospheric dynamics, as well as address potential implications of uncertainty in fire behavior and the associated input parameters.

4.1. Vertical Plume Rise in the Boundary Layer

As demonstrated in our results summary in Section 3.2, initially WRF-SFIRE produced a fairly accurate near-source emissions distribution and plume top with a slight under-prediction of concentrations (Figure 5a).

Over time model performance appears to deteriorate. Given that the fire thermal forcing compares relatively well with observations (Section 3.1), a more likely cause for the increasing difference between model and observations is background boundary layer dynamics. The atmosphere was initiated with 10:00 CST sounding, and continually forced with an observations-based constant surface heat flux. However, the cyclic lateral boundary conditions maintained the same vertical wind profile as initially supplied by the sounding at 10:00 CST, irrespective of potentially changing mesoscale conditions in the real atmosphere. Over the course of more than three hours between spin up start and the final minutes of the fire simulation, from which the corkscrew emissions distribution was obtained (Figure 5b), the real atmospheric wind profile likely evolved.

With time and further downwind the effects of any small changes in mesoscale conditions become more pronounced, which is why initially encouraging model performance deteriorated towards the end of the simulation. The markedly narrow band of emissions in Figure 5b suggests that the "corkscrew" location in the LES domain corresponded to the very edge of the plume rather than the center, indicating a shift in mesoscale wind conditions.

Indeed, analysis of observed background 30 m wind direction leading up to and during the burn shows a significant shift to the west, resulting in the LES "corkscrew" profile being extracted from the edge of the plume, rather then the intended center (Figure 6). Accounting for this observed wind rotation, it is possible to extract a wind-corrected profile, such as shown with a red dotted line in Figure 5b. Assuming an average 20 degree rotation over the course of available wind observations (based on the slope of linear regression shown Figure 6a), the corrected location of the corkscrew maneuver indeed corresponds to the center of the plume (Figure 6b). The wind-corrected profile shown in Figure 5b is a notable improvement from the original non-rotated estimate. Note that this adjustment is extremely crude, as it is based on an estimated wind rotation at one point on a single vertical level and does not take into account potential changes in vertical wind shear.

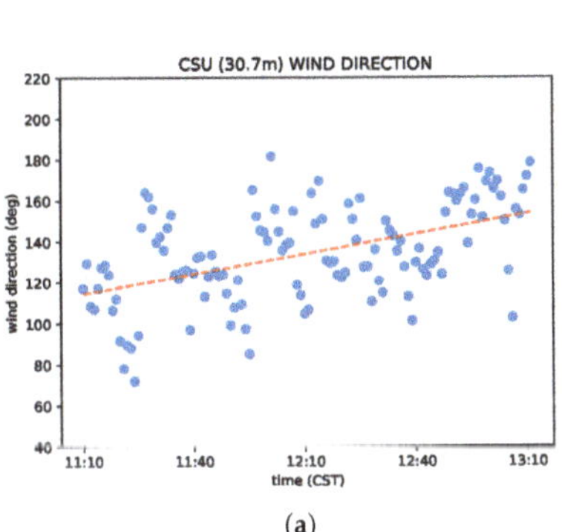

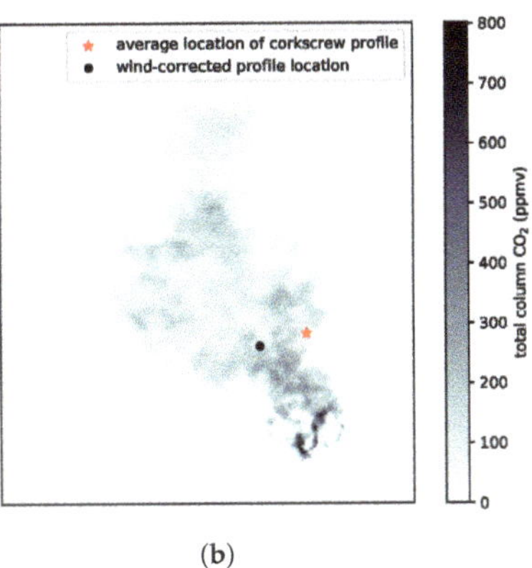

(a)   (b)

Figure 6. The effects of changing mesoscale wind conditions on plume observations (a) Observed change in 30 m wind direction prior to and during the burn. Significant linear trend is shown with a red dashed line. (b) Top view of modeled smoke plume during the "corkscrew" maneuver by the instrumented aircraft. Black dot and red star indicate the average location of the "corkscrew" profile from flight with and without wind-correction, respectively.

Unfortunately, unlike the Real-mode WRF simulations, there is no easy way to account for changing lateral boundary conditions in WRF-SFIRE large-eddy mode. Hence, we can expect the ability of the model to accurately capture dispersion to depend strongly on the variability of real background conditions as well as the simulation length and spatial extent of the modeled domain. Namely, an LES will provide better simulations for situations where that actual atmosphere is horizontally uniform and temporally steady. While this presents a limitation for smoke plume rise and dispersion modelers, it is important to consider it in the context of existing alternative sources of field data. Given a typical uncertainty of ~500 m associated with the most accurate widely available plume height dataset from Multi-angle Image SpectroRadiometer (MISR) [22], WRF-SFIRE provides a valuable alternative source for generating comparatively accurate "synthetic plume height data".

Moreover, unlike instantaneous observational point measurements or overpass-limited derived satellite data, the LES allows us to examine the domain-wide temporal evolution of the plume and identify key features, which are likely to be of interest to dispersion modelers. As shown in Figure 7 and Animation S2, the vertical distribution of emissions in the domain changes throughout the simulation. Following an initial overshoot and a period of active smoke production near the ground, most of the emissions rise and end up near the top of the BL, accumulating just under the inversion level in a wide span of heights. While this vertical distribution may contain modeling and initial condition biases, it is likely to offer dispersion modelers an advantage over the common current approach of using a single empirically derived injection height.

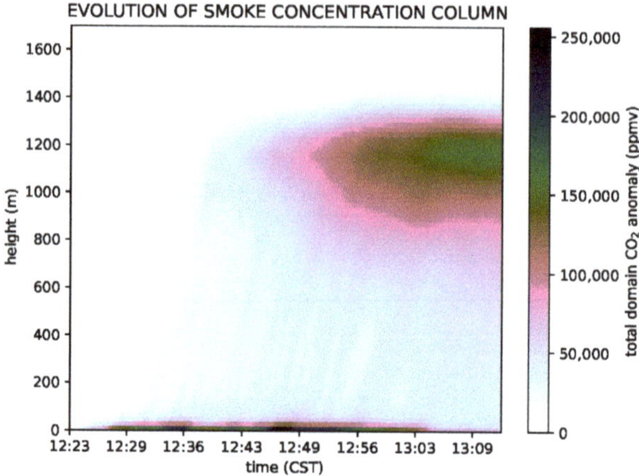

Figure 7. Evolution of total column CO_2 anomaly.

4.2. Importance of Fire Input Parameters

As noted in the Introduction, our evaluation work focused on assessing the relationship between coupled surface forcing and the atmosphere in WRF-SFIRE rather than on fire behavior. However, as we attempt to move forward from simple uncoupled burner-type experiments with prescribed constant surface heat flux to more realistic dynamic simulations, we must address the challenges in selecting proper fire input parameters.

Similar to Kochanski et al. [5], we found that the fire behavior model is particularly sensitive to the choice of fuel moisture. This parameter in WRF-SFIRE does not depend on the selected fuel category and was based entirely on measurements in our simulation. We also modified the standard fuel depth and loading parameters associated with Category 1 fuels to match observations, which resulted in very accurate surface heat flux forcing but substantially lower ROS values than observed or those obtained with standard settings.

Notably, similar thermal forcing to the atmosphere can be produced using a range of combinations of fuel categories and parameters in the model. We have not carried out a formal sensitivity analysis as it was beyond our scope and computational abilities, however, future modelers may find the following information helpful. As preliminary tests for our study, we have used Category 1 and Category 3 fuels (short and tall grass) with various combinations of both standard and measurement-based fuel depth and loading parameters to achieve similar surface forcing. The relationships between these parameters are highly non-linear, which makes determining the "correct" choice (in the absence of detailed observational data) difficult. What we found to be encouraging is that while the absolute value of modeled concentrations and ROS changes dramatically depending on the chosen fuel category for a given fire intensity, the relative distribution of emissions does not. The simulated atmosphere is forced solely by the parameterized heat and moisture fluxes, so WRF-SFIRE does not discriminate which combination of fuel characteristics produced a given heat flux that drives the buoyant plume rise.

Given any thermal forcing, the atmospheric response appears to be fairly robust, irrespective of the particular combination of fuel parameters or ROS with which it was achieved. While this study does not aim to establish whether the model sensitivity to fuel conditions is physical, it does suggest that the LES produces realistic plume rise for the given fire intensity.

4.3. ROS and Biases in Modeled Emissions

The model's poor performance for ROS in our case study likely resulted in reduced simulated emissions concentrations due to lower parameterized fuel consumption rate. This is consistent with the notable negative bias in our modeled CO_2 profiles.

As mentioned above, the low ROS values on our simulation are largely a result of our use of non-standard fuel depth and loading parameters. To eliminate alternative causes for slow fire line advance, we compared horizontal winds at the first and second model levels (at $\sim$8 m and $\sim$25 m AGL) with data obtained from 2D sonic anemometers mounted at multiple heights of the CSU-MAPS meteorological tower. As shown in Figure 8, the near-surface winds are generally accurately captured by the model. At the lowest vertical level, there tends to be a slight positive bias, which one would expect to contribute to higher rather than lower ROS values.

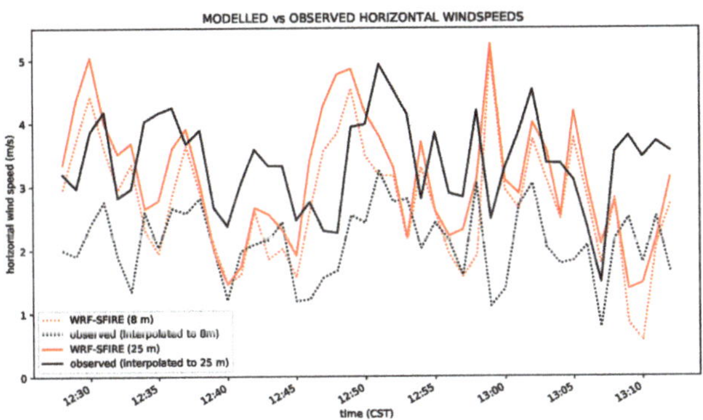

Figure 8. Modeled (red) and observed (black) near-surface horizontal wind.

Apart from their dependency on ROS and fuel consumption, the absolute values of WRF-SFIRE emissions are also controlled by user-prescribed emission factors. In our case study, these factors were not derived from measurements, but were rather based on standard values typical for the Grassland fuel category (see Section 2.2). Hence, the negative bias in our modeled smoke distribution could potentially be reduced, should observations-based emissions factors become available.

4.4. Experimental Design Considerations

One of the shortcomings of the RxCADRE dataset and this experiment is the substantial (nearly 2.5 h) difference in timing between the sounding balloon launch and the fire ignition. Availability of an additional vertical profile for model evaluation just prior to ignition would have been extremely helpful in mitigating some of the sources of error mentioned in the above sections. A similar recommendation was offered by Kochanski et al. [4], who suggested that an on-site sounding just prior to the burn rather than a few hours earlier would be most useful.

While we recognize the challenges of coordinating balloon launches in the presence of aircraft over the fire, a potential alternative would be to include on-board temperature and wind sensor data from flight with the smoke dispersion measurements.

4.5. Limitations

Recent studies suggest that the heat extinction depth parameter in WRF-SFIRE (or e-folding distance) has a strong influence on the modeled fire and near surface plume behavior [4,23]. Currently, there is no clear theory on how the vertical distribution of fire-released heat above the ground affects near-ground temperatures as well as ROS in the literature. As the relationship appears to be highly non-linear, we have not examined its implications in our simulations.

Overall, our findings suggest that the ability of WRF-SFIRE to capture plume dynamics of a specific real fire largely depends on the availability of timely atmospheric initial conditions and accurate simulation of fire intensity. Owing to the detail and comprehensive nature of the data provided by the RxCADRE experiment, these critical inputs could generally be derived from measurements for the current case study. This sensitivity, however, could present a challenge for future real-time fire simulations, where few or no such measurements would be available.

5. Conclusions

This work aimed to assess the ability of a coupled fire–atmosphere WRF-SFIRE LES model to simulate a case study of fire smoke plume growth and dispersion. We examined the L2G burn from the RxCADRE 2012 campaign—a comprehensive experiment combining simultaneous monitoring of fuel, fire behavior, meteorology and emissions.

Our model evaluation demonstrates good overall agreement between the LES and the observations, subject to accuracy and timeliness of model initialization data. Using the emissions and dispersion data collected from an airborne platform during the RxCADRE experiment, we show that LES reasonably captures the timing, rise and dispersion of the fire plume. We examined the possible relationships among model biases, fire behavior and changes in ambient atmospheric conditions.

The work demonstrates the utility of WRF-SFIRE LES in studying some aspects of fire plume dynamics. The scarcity of detailed plume observations presents one of the central challenges for smoke-model development. WRF-SFIRE's ability to capture the rise and spread of fire emissions for cases such as studied here has the potential to address this critical research need and provide alternative "synthetic" data for future development of parameterizations for wildfire smoke plume rise.

Supplementary Materials: The following are available online at http://www.mdpi.com/2073-4433/10/10/579/s1, Animation S1: WRF-SFIRE simulated fire and smoke over real terrain. Visualization produced using VAPOR software. Animation S2: Cross section of WRF-SFIRE simulated emissions along mean wind direction. WRF-SFIRE_init_files.zip: All files required to initialize and run the model simulation.

Author Contributions: Conceptualization, N.M. and R.S.; Formal analysis, N.M.; Funding acquisition, R.S.; Methodology, N.M.; Resources, R.S.; Supervision, R.S.; Visualization, N.M.; Writing—original draft, N.M.; and Writing—review and editing, R.S.

Funding: This work was funded by grants from Natural Sciences and Engineering Research Council of Canada (NSERC) and BC Clean Air Research Fund (CLEAR).

Acknowledgments: The authors would like to acknowledge Brian Potter, Ronan Paugam, Ruddy Mell, Derek McNaran and Adam Kochanski for their input and collaboration. Thanks are also given to Daisuke Seto and Craig Clements for their help with obtaining RxCADRE sounding data and the UBC Weather Research and Forecasting Team for their ongoing support.

Conflicts of Interest: The authors declare no conflict of interest.

References

1. Coen, J.L.; Cameron, M.; Michalakes, J.; Patton, E.G.; Riggan, P.J.; Yedinak, K.M. WRF-Fire: Coupled Weather—Wildland Fire Modeling with the Weather Research and Forecasting Model. *J. Appl. Meteorol. Climatol.* **2012**, *52*, 16–38. [CrossRef]

2. Mandel, J.; Beezley, J.D.; Kochanski, A.K. Coupled atmosphere-wildland fire modeling with WRF 3.3 and SFIRE 2011. *Geosci. Model Dev.* **2011**, *4*, 591–610. [CrossRef]

3. Mandel, J.; Amram, S.; Beezley, J.; Kelman, G.; Kochanski, A.; Kondratenko, V.; Lynn, B.; Regev, B.; Vejmelka, M. Recent advances and applications of WRF–SFIRE. *Nat. Hazards Earth Syst. Sci.* **2014**, *14*, 2829–2845. [CrossRef]

4. Kochanski, A.K.; Jenkins, M.A.; Mandel, J.; Beezley, J.D.; Clements, C.B.; Krueger, S. Evaluation of WRF-SFIRE performance with field observations from the FireFlux experiment. *Geosci. Model Dev.* **2013**, *6*, 1109–1126. [CrossRef]

5. Kochanski, A.; Jenkins, M.; Mandel, J.; Beezley, J.; Krueger, S. Real time simulation of 2007 Santa Ana fires. *For. Ecol. Manag.* **2013**, *294*, 136–149. [CrossRef]

6.   Kochanski, A.K.; Jenkins, M.A.; Yedinak, K.; Mandel, J.; Beezley, J.; Lamb, B. Toward an integrated system for fire, smoke and air quality simulations. *Int. J. Wildland Fire* **2016**, *25*, 534–546. [CrossRef]

7.   Mallia, D.; Kochanski, A.; Urbanski, S.; Lin, J. Optimizing smoke and plume rise modeling approaches at local scales. *Atmosphere* **2018**, *9*, 166. [CrossRef]

8.   Ottmar, R.D.; Hiers, J.K.; Butler, B.W.; Clements, C.B.; Dickinson, M.B.; Hudak, A.T.; O'Brien, J.J.; Potter, B.E.; Rowell, E.M.; Strand, T.M.; et al. Measurements, datasets and preliminary results from the RxCADRE project—2008, 2011 and 2012. *Int. J. Wildland Fire* **2016**, *25*, 1–9. [CrossRef]

9.   Clements, C.B.; Lareau, N.P.; Seto, D.; Contezac, J.; Davis, B.; Teske, C.; Zajkowski, T.J.; Hudak, A.T.; Bright, B.C.; Dickinson, M.B.; et al. Fire weather conditions and fire–atmosphere interactions observed during low-intensity prescribed fires–RxCADRE 2012. *Int. J. Wildland Fire* **2016**, *25*, 90–101. [CrossRef]

10.   Butler, B.; Teske, C.; Jimenez, D.; O'Brien, J.; Sopko, P.; Wold, C.; Vosburgh, M.; Hornsby, B.; Loudermilk, E. Observations of energy transport and rate of spreads from low-intensity fires in longleaf pine habitat–RxCADRE 2012. *Int. J. Wildland Fire* **2016**, *25*, 76–89. [CrossRef]

11.   Hudak, A.T.; Bright, B.C. RxCADRE 2008, 2011, and 2012: Burn Blocks. 2014. Available online: https://doi.org/10.2737/RDS-2014-0031 (accessed on 16 July 2019).

12.   Hudak, A.T.; Bright, B.C.; Kremens, R.L.; Dickinson, M.B. RxCADRE 2011 and 2012: Wildfire Airborne Sensor Program Long Wave Infrared Calibrated Image Mosaics. 2015. Available online: https://doi.org/10.2737/RDS-2016-0008 (accessed on 16 July 2019).

13.   Seto, D.; Clements, C.B. RxCADRE 2012: CSU-MAPS Wind LiDAR Velocity and Microwave Temperature/Relative Humidity Profiler Data. 2015. Available online: https://doi.org/10.2737/RDS-2015-0026 (accessed on 16 July 2019).

14.   Seto, D.; Clements, C.B. RxCADRE 2012: In-Situ Wind, Air Temperature, Barometric Pressure, and Heat Flux Time Series Data. 2015. Available online: https://doi.org/10.2737/RDS-2015-0048 (accessed on 16 July 2019).

15.   Hudak, A.T.; Bright, B.C.; Williams, B.W.; Hiers, J.K. RxCADRE 2011 and 2012: Ignition Data. 2017. Available online: https://doi.org/10.2737/RDS-2017-0065 (accessed on 16 July 2019).

16.   Ottmar, R.D.; Restaino, J.C. RxCADRE 2008, 2011, and 2012: Ground Fuel Measurements from Prescribed Fires. 2014. Available online: https://doi.org/10.2737/RDS-2014-0028 (accessed on 16 July 2019).

17.   Urbanski, S.P. RxCADRE 2012: Airborne Measurements of Smoke Emission and Dispersion from Prescribed Fires. 2014. Available online: https://doi.org/10.2737/RDS-2014-0015 (accessed on 16 July 2019).

18.   Kartsios, S.; Karacostas, T.; Pytharoulis, I.; Dimitrakopoulos, A. Coupled Weather-Wildland Fire Model for fire behavior interpretation. In Proceedings of the 12th International Conference on Meteorology, Climatology and Atmospheric Physics, Heraklion, Greece, 28–31 May 2014.

19.   Anderson, H.E. *Aids to Determining Fuel Models for Estimating Fire Behavior*; General Technical Report INT-122; U.S. Department of Agriculture: Washington, DC, USA, 1982; p. 22.

20.   Overholt, K.; Cabrera, J.; Kurzawski, A.; Koopersmith, M.; Ezekoye, O. Characterization of fuel properties and fire spread rates for little bluestem grass. *Fire Technol.* **2014**, *50*, 9–38. [CrossRef]

21.   Prichard, S.; O'Neill, S.; Urbanski, S. *Evaluation of Revised Emissions Factors for Emissions Prediction and Smoke Management*; United States Environmental Protection Agency: Washington, DC, USA, 2017.

22.   Val Martin, M.; Kahn, R.; Tosca, M. A Global Analysis of Wildfire Smoke Injection Heights Derived from Space-Based Multi-Angle Imaging. *Remote Sens.* **2018**, *10*, 1609. [CrossRef]

23.   Kartsios, S.; Karacostas, T.S.; Pytharoulis, I.; Dimitrakopoulos, A.P. The Role of Heat Extinction Depth Concept to Fire Behavior: An Application to WRF-SFIRE Model. In *Perspectives on Atmospheric Sciences*; Karacostas, T., Bais, A., Nastos, P.T., Eds.; Springer International Publishing: Cham, Switzerland, 2017; pp. 137–142.

Article

Differences in Model Performance and Source Sensitivities for Sulfate Aerosol Resulting from Updates of the Aqueous- and Gas-Phase Oxidation Pathways for a Winter Pollution Episode in Tokyo, Japan

Syuichi Itahashi [1],*, Kazuyo Yamaji [2], Satoru Chatani [3] and Hiroshi Hayami [1]

[1] Central Research Institute of Electric Power Industry, Abiko, Chiba 270-1194, Japan; haya@criepi.denken.or.jp
[2] Graduate School of Maritime Sciences, Kobe University, Kobe, Hyogo 658-0022, Japan; kazuyo@maritime.kobe-u.ac.jp
[3] National Institute for Environmental Studies, Tsukuba, Ibaraki 305-8506, Japan; chatani.satoru@nies.go.jp
* Correspondence: isyuichi@criepi.denken.or.jp

Received: 31 July 2019; Accepted: 9 September 2019; Published: 12 September 2019

Abstract: During the Japanese intercomparison study, Japan's Study for Reference Air Quality Modeling (J-STREAM), it was found that wintertime SO_4^{2-} concentrations were underestimated over Japan with the Community Multiscale Air Quality (CMAQ) modeling system. Previously, following two development phases, model performance was improved by refining the Fe- and Mn-catalyzed oxidation pathways and by including an additional aqueous-phase pathway via NO_2 oxidation. In a third phase, we examined a winter haze period in December 2016, involving a gas-phase oxidation pathway whereby three stabilized Criegee intermediates (SCI) were incorporated into the model. We also included options for a kinetic mass transfer aqueous-phase calculation. According to statistical analysis, simulations compared well with hourly SO_4^{2-} observations in Tokyo. Source sensitivities for four domestic emission sources (transportation, stationary combustion, fugitive VOC, and agricultural NH_3) were investigated. During the haze period, contributions from other sources (overseas and volcanic emissions) dominated, while domestic sources, including transportation and fuel combustion, played a role in enhancing SO_4^{2-} concentrations around Tokyo Bay. Updating the aqueous phase metal catalyzed and NO_2 oxidation pathways lead to increase contribution from other sources, and the additional gas phase SCI chemistry provided a link between fugitive VOC emission and SO_4^{2-} concentration via changes in O_3 concentration.

Keywords: Community Multiscale Air Quality (CMAQ); East Asia; Tokyo; SO_4^{2-}; stabilized Criegee intermediates (SCI)

1. Introduction

To improve our understanding of the behavior of air pollutants, advances in three-dimensional air quality modeling are necessary. Modeling systems can represent key environmental processes (emission, transport, chemical reactions, deposition) which determine the behavior of air pollutants; however, there are uncertainties in these processes. To better understand these uncertainties and improve modeling performance, intercomparison studies can be of great value. Based on the results and experience of projects in Japan, an intercomparison project called Japan's Study for Reference Air Quality Modeling (J-STREAM) was initiated [1], and this has provided insights into how to improve modeling [2–5]. J-STREAM aims to establish reference air quality models for source apportionment and to formulate a strategy for mitigating air pollutants in Japan, including particulate matter with diameters less than 2.5 μm (PM2.5) and photochemical ozone (O_3). The first phase of J-STREAM focused on understanding the ranges and limitations of PM2.5 and O_3 concentrations simulated by participants using common input datasets. Simulations for the first phase were conducted over two weeks in each season from January 2013 to March 2014 in accordance with official monitoring programs for PM2.5 [1]. In Japan, sulfate (SO_4^{2-}) aerosol is a major component of PM2.5 and it was found that models generally capture the observations but underestimate the concentrations in winter. Therefore, the model processes that were related to SO_4^{2-} production were carefully reviewed, revealing inadequate oxidation of SO_2 via aqueous-phase reactions of O_2 catalyzed by Fe and Mn because of a lack of trace metal data over Asia in the current emission inventory [3]. For the second phase of J-STREAM, a winter haze episode for the period 13–25 December 2016 was targeted. In addition to the incorporation of refinements for the Fe- and Mn-catalyzed aqueous oxidation in the first phase, the aqueous-phase reaction via NO_2 was introduced, and this increased the SO_4^{2-} production levels to catch the SO_4^{2-} observations. Intercomparison studies involving two state-of-the-art regional models, the Community Multiscale Air Quality Model (CMAQ) and the Comprehensive Air Quality Model with eXtensions (CAMx) were also conducted. As a result, it was suggested that differences in model performances were possibly [4] caused by wet deposition processes. Thus, a third phase of J-STREAM, which is the focus of the present study, was conducted, which again involved the winter haze episode examined in second phase. The main purpose of third phase was to obtain and compare the source sensitivities with respect to major emission sources by model intercomparison.

The remainder of this paper is organized as follows. The model setups for CMAQ are described in the next section. To improve the estimations for the ambient concentrations of SO_4^{2-}, an additional gas-phase oxidation pathway involving stabilized Criegee intermediates (SCI) was included in the model and full details of the reaction chemistry are given. In the results and discussion section, model performances were evaluated for observations in Tokyo together with an analysis of source sensitivities. The conclusions are presented in the final section.

2. Modeling Design

The third phase of J-STREAM focuses again on the winter haze episode of 15–25 December 2016 [4]. The domain and the common input dataset for the emissions and the meteorology basically followed that for the J-STREAM framework [1]. Domain 1 covered the whole of Asia, domain 2 covered the whole of Japan and domains 3 and 4 covered the Kansai region (including Osaka and Nagoya) and the Kanto region (including Tokyo), respectively. The horizontal grid resolution was 45 km for domain 1, 15 km for domain 2, and 5 km for domains 3 and 4. In the second phase, the vegetation database for Japan was introduced and this revision helped to improve the meteorological fields and the emissions of biogenic volatile organic compounds [2]. For the emissions inventory, the compositions of metal elements in PM2.5 over Asia were considered in accordance with chemical composition reports [6] based on our recommendations in first phase [4]. In the third phase, anthropogenic emissions from China were revised by adjusting data for 2016 that included updated emission information. For example, the SO_2 emissions from China in 2016 was almost half of emissions in 2010 [7]. For meteorology the revision was performed in this third phase on the configuration of Weather Research and Forecasting (WRF) version 3.7.1 [8]. The revised points are summarized as follows and others are based on the established J-STREAM framework in the first phase. The top pressure level was set on 50 hPa. The reanalysis data was taken from the National Centers for Environmental Prediction (NCEP) final analysis (ds083.3) with 6-h intervals and with a 0.25° horizontal grid resolution [9], and the sea surface temperature (SST) data was obtained from the Group for High Resolution Sea Surface Temperature (GHRSST) of level 4 with 24-h intervals and with 1-km horizontal grid resolution [10] for the initial and boundary conditions. The grid nudging on wind was conducted on all layers, and those on temperature and water vapor were not conducted in the planetary boundary layer (PBL). The nudging coefficient for wind was 1.0×10^{-4} for all domains and those for temperature and water vapor were 5.0×10^{-5} for domain 1, 3.0×10^{-5} for domain 2, and 1.0×10^{-5} for domains 3 and 4. The shortwave and longwave radiation scheme used RRTMG [11], and the microphysics and cumulus options respectively adopted the Morrison double-moment scheme [12] and the Grell–Devenyi ensemble scheme [13] based on the performance improvement on pre-simulation.

CMAQ version 5.2 [14] model simulations were performed in this study. The gas and aerosol chemistry were handled by SAPRC07 [15] and AERO6, respectively. In connection with SO_4^{2-} production, CMAQ version 5.2 features one gas-phase chemical reaction and five aqueous-phase chemical reactions [16]. The original configuration released as CMAQ (hereafter referred to as the base-case simulation) was simulated first. Then, the simulation of chemistry updates A was performed based on our findings in the first phase [3]. The aqueous-phase oxidation pathways for O_2 via Fe and Mn catalysis were refined by increasing the anthropogenic Fe and Mn solubilities from 10% to 25% and from 50% to 100%, respectively. In this aqueous-phase oxidation pathway for O_2:

$$-d[S(IV)]/dt = k_1[Fe(III)][S(IV)] + k_2[Mn(II)][S(IV)] + k_3[Fe(III)][Mn(II)][S(IV)] \qquad (1)$$

considered the pH dependency of the rate constants as follows when the synergistic existence of Fe and Mn,

$$k_3'[H^+]^{0.67}[Fe(III)][Mn(II)][S(IV)] \ (pH \geq 4.2),$$
$$k_3''[H^+]^{-0.74}[Fe(III)][Mn(II)][S(IV)] \ (pH < 4.2), \qquad (2)$$

where $k_3' = 2.51 \times 10^{13} \ M^{-1} s^{-1}$ and $k_3'' = 3.72 \times 10^7 \ M^{-1} s^{-1}$ [3]. Moreover, the aqueous-phase reaction pathway via NO_2 was introduced in CMAQ:

$$-d[S(IV)]/dt = k[NO_2(aq)][S(IV)], \qquad (3)$$

based on consideration of the neutralized or acidic features of aerosols in Asia and having a rate constant expression as follows:

$$k = 1.24 \times 10^7 \ M^{-1} s^{-1} \ (pH < 5.3), \ k = 1.67 \times 10^7 \ M^{-1} s^{-1} \ (pH > 8.7), \qquad (4)$$

where for the pH range 5.3–8.7, the rate constant was linearly interpolated based on our findings in the second phase [4]. These revisions, which focused on aqueous-phase oxidation pathways, were included in chemistry updates A.

Previous studies [3,4] have focused on the need to refine the aqueous-phase sulfur oxidation pathways, but not the gas-phase reactions. Although refinements of the aqueous-phase oxidation pathways resulted in an improvement in model performance, underestimations of SO_4^{2-} concentrations were not corrected. In addition, the increase in SO_4^{2-} production via aqueous-phase oxidation also led to an increase in SO_4^{2-} wet deposition. Regarding the relation between ambient concentrations of SO_4^{2-} and deposition of SO_4^{2-}, a comparison between CMAQ and CAMx with observation demonstrated the potential for overestimation of SO_4^{2-} wet deposition by CMAQ [4]. Originally, the CMAQ model considered one gas-phase reaction of SO_2 to be oxidized by an OH radical. Possible pathways of gas-phase oxidation involve SCI, which are produced from the reaction of alkenes and O_3 [17,18]. A review of the SCI rate constants pointed out the wide range of values, covering three orders of magnitude [19]. This was due to the lack of direct measurement techniques available at the time to detect SCI. The impact of the simplest SCI of formaldehyde oxide (CH_2OO) has been examined based on a recently measured rate constant [20]. Also, direct measurement of CH_2OO has been reported by another research group who found a similarly elevated rate constant for SO_4^{2-} production, but a different rate constant for H_2O [21]. The reactions of CH_2OO with methanol (CH_3OH), ethanol (CH_3CH_2OH), and 2–propanol (($CH_3)_2CHOH$) have also been reported [22]. In addition, the higher SCI of acetaldehyde oxide (CH_3CHOO) and propionaldehyde oxide (($CH_3)_2COO$) have also been reported recently [23,24]. A study on the application of CMAQ in the U.S.A., including the use of the representative SCI gas-phase oxidation into an updated 2005 Carbon Bond (CB05), reported the potential impacts on SO_4^{2-} production [25]. Another study investigated the role of SCI based on the Master Chemical Mechanism (MCM) [26]. In this study, as a result of advances made in direct measurement of SCI, the gas-phase chemistry of three SCI (CH_2OO, SCI1; CH_3CHOO, SCI2; $(CH_3)_2COO$, SCI3) has been explicitly incorporated in SAPRC07. The yields of these three SCI species are derived directly from the yields of corresponded formic, acetic, and propanoic acids used in the gas-phase chemical reactions of SAPRC07 in CMAQ. The revised and added reactions are summarized in Table 1. The rate expression for SCI1 to H_2O posed a potential issue for SO_4^{2-} production given that previous studies concluded that a higher rate constant for the reaction of SCI and H_2O led to the consumption of SCI by H_2O [25,26]. In the case of SCI2, this specie has geometric isomers (the syn- and anti- forms) which exhibit different rate coefficients [23]; however, we simply averaged these values, treating SCI2 as a single entity. Thus, in addition to the revision of the aqueous-phase reactions in chemistry updates A, these gas-phase oxidation pathways involving the three SCI species were considered in the chemistry updates B.

Table 1. Reactions involving the three stabilized Criegee intermediates (SCI) species on SAPRC07 adapted in this study.

Reaction	Rate Constant	Reference
$O_3 + ETHE \rightarrow \ldots + 0.370 \times SCI1$		[15]
$O_3 + PRPE \rightarrow \ldots + 0.185 \times SCI1 + 0.075 \times SCI2$		[15]
$O_3 + BD13 \rightarrow \ldots + 0.185 \times SCI1$		[15]
$O_3 + OLE1 \rightarrow \ldots + 0.185 \times SCI1 + 0.159 \times SCI3$		[15]
$O_3 + OLE2 \rightarrow \ldots + 0.024 \times SCI1 + 0.065 \times SCI2 + 0.235 \times SCI3$		[15]
$O_3 + ISOP \rightarrow \ldots + 0.204 \times SCI1$		[15]
$O_3 + IPRD \rightarrow \ldots + 0.100 \times SCI1 + 0.372 \times SCI3$		[15]
$O_3 + TERP \rightarrow \ldots + 0.172 \times SCI1 + 0.068 \times SCI3$		[15]
$O_3 + SESQ \rightarrow \ldots + 0.172 \times SCI1 + 0.058 \times SCI3$		[15]
$SCI1 + SO_2 \rightarrow HCHO + SULF$	3.9×10^{-11}	[20]
$SCI1 + NO_2 \rightarrow HCHO + NO_3$	1.5×10^{-12}	[21]
$SCI1 + NO \rightarrow HCHO + NO_2$	2.0×10^{-13}	[21]
$SCI1 + H_2O \rightarrow$	2.4×10^{-15}	[20]
	9.0×10^{-17}	[21]
$SCI1 + MEOH \rightarrow$	1.4×10^{-13}	[22]
$SCI1 + ETOH \rightarrow$	2.3×10^{-13}	[22]
$SCI1 + ALK4 \rightarrow$	1.9×10^{-13}	[22]
$SCI2 + SO_2 \rightarrow CCHO + SULF$	4.55×10^{-11}	[23]
$SCI2 + H_2O \rightarrow$	7.0×10^{-14}	[23]
$SCI3 + SO_2 \rightarrow RCHO + SULF$	1.3×10^{-10}	[24]
$SCI3 + H_2O \rightarrow$	1.5×10^{-16}	[24]

Note: Unit of rate constant is $cm^3 \ s^{-1}$. The names are based on the nomenclature used in the expressions in SAPRC07; O_3, ozone; ETHE, ethene; PRPE, propene; BD13, 1,3-butadiene; OLE1 refers to alkenes with reaction rates with $OH < 7.0 \times 10^{-4}$ ppm^{-1} min^{-1} (excluding ethene); OLE2 refers to alkenes with reaction rates with $OH > 7.0 \times 10^{-4}$ ppm^{-1} min^{-1}; ISOP, isoprene; IPRD, lumped isoprene products; TERP, terpene; SESQ, sesquiterpenes; SO_2, sulfur dioxide; SULF, sulfate (SO_3 or H_2SO_4); NO_2, nitrogen dioxide; NO_3, nitrate radical; NO, nitric oxide; H_2O, water; MEOH, methanol; ETOH, ethanol; ALK4 refers to alkanes and other non-aromatic compounds that react only with OH with a rate constant range between 5.0×10^{-3} and 1.0×10^{-4} ppm^{-1} min^{-1}; CCHO, acetaldehyde; RCHO, lumped aldehydes; SCI species: SCI1 refers to CH_2OO, SCI2 refers to CH_3CHOO and SCI3 refers to $(CH_3)_2COO$.

The other approach for modeling the aqueous-phase reaction chemistry is the recently developed kinetic mass transfer (KMT) simulation for gas- and aqueous-phase species that simultaneously integrates phase transfer, scavenging, deposition, dissociation, and chemical kinetic processes (AQCHEM-KMT) [27,28] using the Kinetic PreProcessor [29]. It was reported that there was no significant impact on the monthly averaged data, but possible differences at the hourly timescale were noted [27]. To test this newly developed approach, a KMT simulation was also conducted. The simulations performed in this study are summarized in Table 2.

Table 2. Summary of Community Multiscale Air Quality (CMAQ) simulations conducted in this study.

Name	Description
Chemistry Updates A	Fe and Mn solubilities are increased and the rate constant expression for the Fe- and Mn-catalyzed oxidation by O_2 includes a pH dependency. Addition of an NO_2 aqueous-phase reaction (a total of six aqueous-phase reactions were treated).
Chemistry Updates B	Same as sensitivity Simulation A, but with addition of gas-phase oxidation pathways related to SCI (see Table 1).
Kinetic Mass Transfer (KMT)	Selection of the AQCHEM-KMT option

3. Results and Discussion

3.1. Model Performance

To gain an appreciation of the four simulations performed by CMAQ, the spatial distributions of SO_4^{2-} concentrations simulated in domain 1 of J-STREAM are shown in Figure 1, based on averaging of the entire monitoring period. The high concentrations of SO_4^{2-} over the Asian continent that included the downwind region of Japan were related to transboundary SO_4^{2-} in Japan as discussed previously [30–34]. The differences in distributions of SO_4^{2-} between the base-case simulations and those for the chemistry updates A and KMT are clearly evident. An increase in SO_4^{2-} concentrations of greater than 5% for chemistry updates A compared with the base-case simulation was found for the Korean Peninsula and Japan which is the downwind region of the Asian continent. This finding was the same as demonstrated in our previous study [4]. However, when comparing KMT and the base-case simulation a negative effect was noted, the largest change in SO_4^{2-} concentrations of more than −5% being detected over the Sea of Okhotsk. This region corresponded to a SO_4^{2-} concentration of less than 1.0 μg/m^3 with the absolute change in concentration being less than −0.1 μg/m^3. This result suggests nonsignificant impacts by KMT on averaged SO_4^2 and was consistent with the report on the KMT test case over the U.S.A. [27].

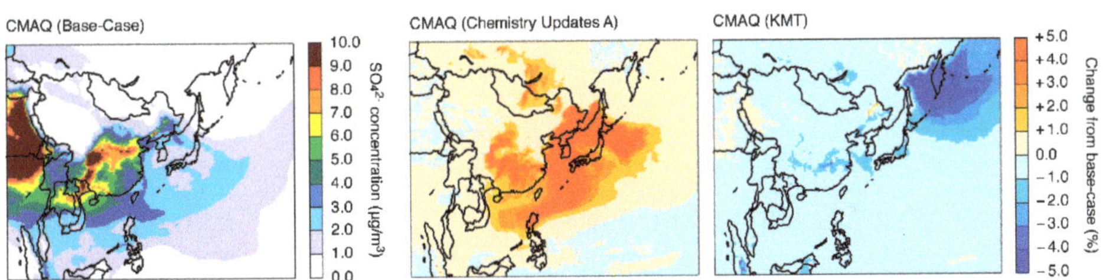

Figure 1. Spatial distribution of (**left**) SO_4^{2-} concentrations simulated by the CMAQ base-case simulation for domain 1 and (**right**) changes in SO_4^{2-} concentrations for chemistry updates A and KMT averaged over the period 15–25 December 2016.

To investigate the effects of including the three SCI in the simulations, each SCI treated in this study was tested in an incremental manner. The results are shown in Figure 2 as the difference from the chemistry updates A. A total of six cases were tested: SCI1 with either a higher or lower rate constant for the reaction of SCI1 with H_2O, SCI1 plus the addition of SCI2, and SCI1 plus SCI2 with the further inclusion of SCI3. The increase in SO_4^{2-} production that occurred with the inclusion of SCI1 was not found in the simulation with the higher rate constant for SCI1 + H_2O but was clearly observed in the simulation with the lower rate constant over mainland China and toward the downwind region of northern Japan, over India and extending to the Bay of Bengal, Bangladesh, and Myanmar, and over some parts of Indonesia. The importance of the rate constant for H_2O which can consume SCI1, as demonstrated over Asia in this study, was suggested in research conducted over the U.S.A. [25].

The addition of SCI2 did not cause an increase in SO_4^{2-} concentrations and this may well reflect the fact that only two chemical reactions were involved in producing SCI2 whose stochastic coefficients were smaller than those for SCI1 (Table 1). The further inclusion of SCI3 led to an increase in SO_4^{2-} concentrations in spite of the use of a higher or a lower rate constant for the reaction of SCI1 with H_2O. Chemistry updates B showed a 1–2% increase in SO_4^{2-} concentrations over mainland China and a 1–3% increase over Northern India, even when using a higher rate constant for SCI1 + H_2O, and showed a greater than 5% increase in SO_4^{2-} concentrations over mainland China and a 1–2% increase over the northern part of Japan and the Kansai and Kanto regions when a lower rate constant for SCI1+H_2O was used. Through incremental testing of the three SCI, the importance of the value of the rate constants for SCI1 + H_2O over Asia was demonstrated as was also revealed in the study over the U.S.A., and clearly indicates the need for distinct treatments of the individual SCI given that SCI3 has a potential impact on SO_4^{2-} production independent of that for the rate constant of SCI1 with H_2O.

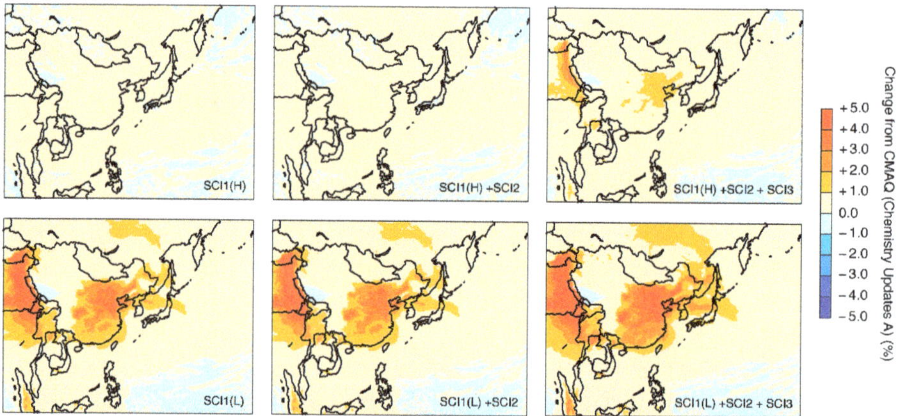

Figure 2. Spatial distribution of SO_4^{2-} concentrations simulated by the CMAQ chemistry updates B as changes from chemistry updates A averaged over the period 15–25 December 2016 in domain 1. The inclusion of SCI was tested incrementally as (**top-left**) only SCI1 with a higher rate constant for H_2O (SCI1(H)), (**top-center**) the addition of SCI2 to SCI1(H), and (**top-right**) the addition of SCI3 to SCI1(H) and SCI2, (**bottom-left**) SCI1 with a lower rate constant of H_2O (SCI1(L)), (**bottom-center**) the addition of SCI2 to SCI1(L), and (**bottom-right**) the addition of SCI3 to SCI(L) and SCI2.

From the incremental testing of SCI, we selected the case for SCI1 with the low rate constant with H_2O, SCI2, and SCI3, hereafter referred to as chemistry updates B. The modeling domain for J-STREAM covered the whole of Japan as domain 2 and the Kanto region (including Tokyo) as domain 4. The simulated ambient SO_4^{2-} concentrations and the wet deposition over domain 4 are shown in Figure 3. High ambient concentrations of SO_4^{2-} were limited to the region over the Tokyo Bay area, and a large amount of wet deposition was found over the southwest areas of the Kanto region. With respect to chemistry updates A and B, the reason for the increase in ambient concentrations of SO_4^{2-} was clarified, whereas changes in the rates of wet deposition for SO_4^{2-} between chemistry updates A and B were not apparent when compared with that for ambient SO_4^{2-} concentrations. As expected, the revision of the gas-phase oxidation pathways, which included the SCI, successfully led to only an increase in ambient SO_4^{2-} concentrations. Using the KMT simulation, slight decreases in both the ambient SO_4^{2-} concentrations and wet deposition were revealed over the Kanto region.

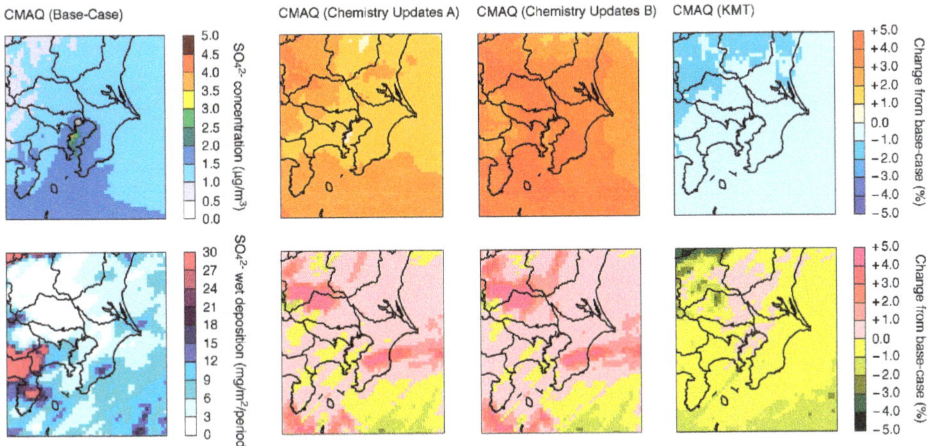

Figure 3. Spatial distribution of (**top**) ambient concentrations of SO_4^{2-} and (**bottom**) wet deposition simulated by (**left**) CMAQ base-case simulation and (**right**) changes by chemistry updates A, B, and KMT relative to the base-case averaged over the period 15–25 December 2016 over domain 4. The observation site at Tokyo is indicated by the gray circle.

A detailed comparison with experimental observations was conducted using an aerosol chemical speciation analyzer (ACSA) automated monitoring system (Kimoto Electric, Co., Ltd., Osaka, Japan) at Mukoujima (139.81° E, 35.71° N) for hourly measurements of ambient concentrations of SO_4^{2-} and the network observations from the Acid Deposition Monitoring Network in East Asia (EANET) at the Tokyo site (139.76° E, 35.69° N) for daily SO_4^{2-} wet deposition (gray circle, Figure 3). This automated system for monitoring ambient concentrations of SO_4^{2-} by ACSA has been evaluated in a previous study conducted in western Japan where the importance of high-resolution monitoring was demonstrated [35]. The wet deposition of aerosols was measured via the use of automated wet-only samplers and the concentration of SO_4^{2-} in precipitation was determined by ion chromatography [36]. The time series of hourly observations and the CMAQ for ambient concentrations of SO_4^{2-} are shown in Figure 4. The temporal variation of the observed ambient concentrations of SO_4^{2-} gave the following variations during the analysis period: around 1 µg/m^3 during 15–18 December; a subsequent increase from 0.6 µg/m^3 to 3.9 µg/m^3 within 1 day on 18 December; a consistently higher concentration of around 4 µg/m^3 from 19 to 22 December; and a subsequent decrease from 3.8 µg/m^3 to 0.6 µg/m^3 within 1 day on 22 December. All of the four CMAQ simulations generally mimicked the observed temporal variations. A detailed discussion of model performance based on statistical analysis is presented later. The hourly precipitation data for the four models and the daily accumulated SO_4^{2-} wet deposition by experimental observation at the Tokyo site over the campaign period is presented in Figure 4. From this data, it is possible to define an increasing SO_4^{2-} concentration period (P1), a relatively stable period (P2), and a decreasing period (P3). From the modeled hourly precipitation results, P3 clearly corresponded to a rainy day. Over Tokyo, there was no observed rain except in the P3 period and this was also shown by the model. The EANET observation at the Tokyo site was conducted on a daily basis (from 9 a.m. to 9 a.m. the next day); the accumulated precipitation during 22–23 December was 15.5 mm, and the modeled result was 15.2 mm. Wet deposition was only observed and modeled during P3; the accumulated observed and modeled SO_4^{2-} wet deposition data were 6.6 mg/m^2 and 5.0 mg/m^2, respectively. The four simulations by CMAQ did not show much difference in wet deposition.

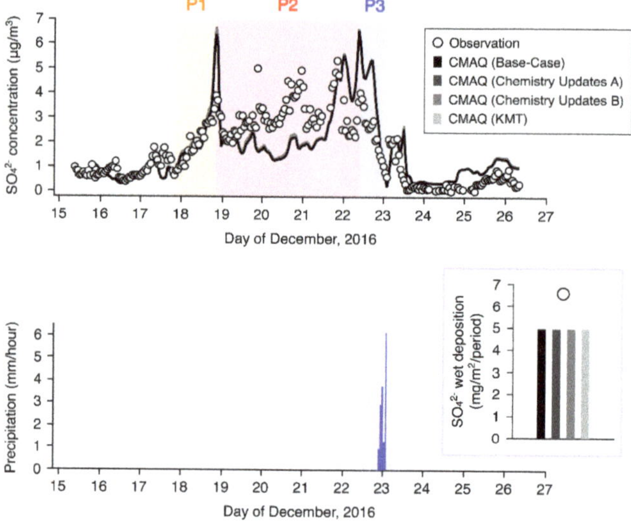

Figure 4. Temporal variation of (**top**) hourly observed and modelled ambient concentrations of $SO_4{}^{2-}$, and (**bottom**) hourly modelled precipitation at the Tokyo site for domain 4. The inset in bottom figure is the daily accumulated $SO_4{}^{2-}$ wet deposition for the observed and modelled results.

For a comparison of the ambient concentrations of $SO_4{}^{2-}$, the model performance was judged according to the statistical analysis data, namely, the correlation coefficient (R) with the significance level determined by the Students' t-test, the normalized mean bias (NMB), the normalized mean error (NME), the mean fractional bias (MFB) and the mean fractional error (MFE). In line with the recommended benchmarks based on the modeling study over the U.S.A., the proposed model performance goals were NMB < ±10%, NME < +35%, and R > 0.70 for the best model performance, and the proposed model performance criteria were NMB < ±30%, NME < +50%, and R > 0.40 for acceptable model performance for the daily $SO_4{}^{2-}$ concentration levels [37]. The model performance goals were also proposed as MFB ≤ ±30% with MFE ≤ +50% for the best model performance, and model performance criteria were proposed as MFB ≤ ±60% with MFE ≤ +75% for acceptable model performance [38]. In addition, the corresponding percentages of the performance terms (within a factor of 2 and 3) were also calculated. The results for statistical analysis are listed in Table 3.

Table 3. Statistical analysis of model performance for $SO_4{}^{2-}$ concentrations at the Tokyo site in domain 4 of J-STREAM.

	Base-Case	Chemistry Updates A	Chemistry Updates B	KMT
N		247		
Mean (observation) [μg/m³]		1.70		
Mean (model) [μg/m³]	1.68	1.70	1.74	1.66
R	0.68 *	0.68 *	0.69 *	0.68 *
	(p < 0.001)	(p < 0.001)	(p < 0.001)	(p < 0.001)
NMB [%]	−1.4 **	0.0 **	+2.6 **	−2.1 **
NME [%]	45.0 *	45.1 *	44.1 *	44.9 *
MFB [%]	+10.7 **	+12.0 **	+14.4 **	+9.8 **
MFE [%]	52.1 *	52.1 *	51.4 *	51.9 *
% within a factor of 2	69.6	69.2	70.5	70.4
% within a factor of 3	87.9	87.9	87.9	88.3

Note: ** indicates model performance goal, and * indicates model performance criteria, see text for these judgements.

Compared with the hourly $SO_4{}^{2-}$ observations observed at Tokyo, all of CMAQ models showed R values of around 0.7 with a statistical significance level of $p < 0.001$. As shown in Figures 1 and 2, chemistry updates A and B led to an increase in $SO_4{}^{2-}$ concentrations whereas KMT led to a slight decrease in $SO_4{}^{2-}$ concentrations. As seen in Figure 4, the differences by KMT were not clarified in hourly time scale in this case applied for the winter pollution episode at Tokyo. Judging from the proposed criteria [37,38], all four CMAQ model simulations were considered to have met the model performance criteria.

The spatial distributions of $SO_4{}^{2-}$ concentrations during the periods P1, P2, and P3 are displayed in Figure 5. The diagram indicates that the region with high $SO_4{}^{2-}$ concentrations greater than 2.0 µg/m^3 in P1 was limited to near the coastline of Tokyo Bay, whereas other areas were below 2.0 µg/m^3 on average. It can be considered that the increased $SO_4{}^{2-}$ concentrations indicated in Figure 3 were not associated with the broad feature spread over the Kanto region but with the limited haze episode over Tokyo. For P2, the high $SO_4{}^{2-}$ concentration regions expanded over the Tokyo Bay area, and nearly the whole of Eastern Kanto (the eastern part of this domain) was covered with high concentrations of $SO_4{}^{2-}$ of around 2.0–2.5 µg/m^3 (dark green color). For P3, the high $SO_4{}^{2-}$ concentrations found over the Tokyo Bay area extended into the northern part of this domain, while the western boundary had lower concentrations of less than 0.5 µg/m^3. The decreased concentrations of $SO_4{}^{2-}$ indicated in Figure 3 would have been be influenced by the intrusion of this low concentration zone. To clarify the reasons for this, a sensitivity analysis against the major domestic sources was conducted.

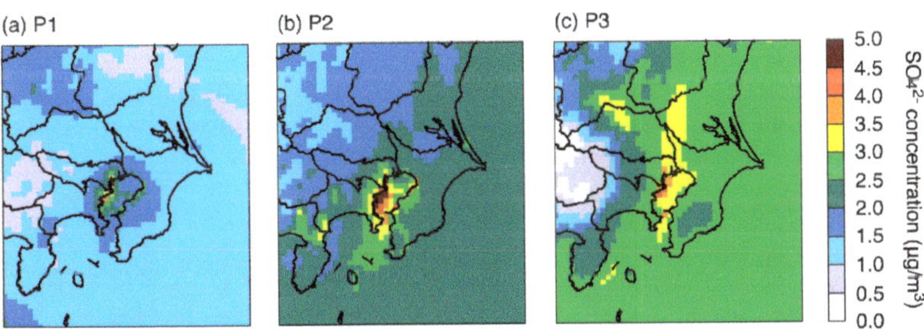

Figure 5. Spatial distribution of $SO_4{}^{2-}$ concentrations simulated by the CMAQ base-case simulation averaged over (**a**) P1, (**b**) P2, and (**c**) P3 for domain 4.

3.2. Model Sensitivities

Here, the sensitivities for major domestic sources were investigated. Four major domestic sources, transportation (automobile, ship, aviation, and machinery), stationary combustion (power plant, industry, waste incinerator), fugitive VOC (fuel evaporation, and solvent use), and agricultural NH_3 (livestock, and fertilizer application). These source groups were numbered g01, g02, g03, and g04, respectively. The SO_2 emissions over domain 4 used in the simulation are shown in Figure 6 as the daily average distributed in two dimensions. High emission levels greater than 500 kg/day were found over the Tokyo Bay area and values greater than 100 kg/day were broadly observed over the central Kanto region and over the ocean. In the sensitivity analysis, SO_2 emissions were contained in g01 and g02 but not in g03 and g04. The emissions from g01 and g02 are also illustrated in Figure 5. It can be clearly seen that SO_2 emissions over the sea were dominated by g01 reflecting ship emissions as being the main source and that emissions over the land were dominated by g02. In this respect, the SO_2 emissions from power plants and industries are mostly centered along the coastline; as a result, SO_2 emissions in the Kanto region were densely concentrated over the Tokyo Bay area.

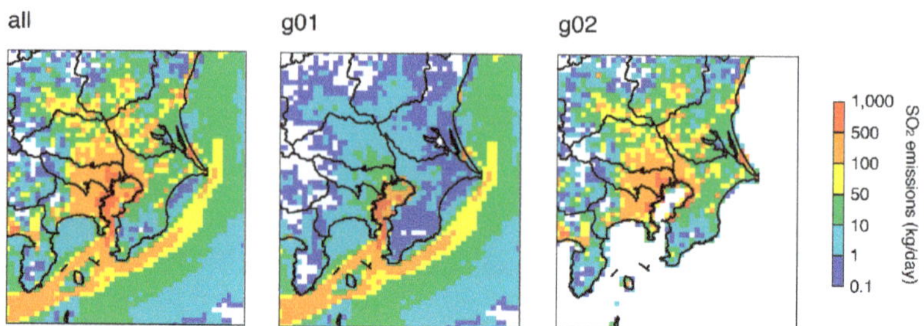

Figure 6. Spatial distribution of (**left**) all SO$_2$ emission sources used in this study and SO$_2$ emissions from sources g01 (**center**) and g02 (**right**) over domain 4. Note that emissions are shown in two dimensions and most of g01 is distributed on the surface layer of the model, whereas g02 is distributed over the upper layer of the model, considering the stack heights.

To obtain the sensitivity values, the traditional so-called "brute force" method was applied. The emission from each source group was reduced by 20% and the simulation was conducted as a sensitivity simulation; next, the difference between the base-case simulation and the sensitivity simulation was calculated. In this study, this difference was multiplied by 5 to correspond to a 100% reduction; accordingly, the source contribution can be regarded as follows:

$$\text{Source contribution of } g_i = (\text{Base-case simulation} - \text{Sensitivity simulation for } g_i) \times 5, \qquad (5)$$

where i = 01, 02, 03, and 04 are the four source groups as defined above. The source contribution of other sources (except the four domestic anthropogenic emissions), such as anthropogenic emissions from outside of Japan, biogenic emissions, biomass burning emissions, and volcanic emissions, are calculated as follows:

$$\text{Source contribution of others} = \text{Base-case simulation} - \sum_{i=01}^{04} \text{Source contribution of } g_i. \qquad (6)$$

A perturbation magnitude of 20% was applied to achieve a compromise between producing a clear signal and applying a sufficiently small perturbation to allow the results to be scaled linearly to a different perturbation level according to the Task Force on Hemispheric Transport of Air Pollution (TF HTAP) modeling [39]. In this study, we focused on four domestic sources and conducted a total of four cases of CMAQ sensitivity simulations on the base-case simulation. In addition to this base-case simulation and the four sensitivity simulations, three other simulations for the chemistry updates A, B, and KMT were also performed against the four emission source groups; as a result, a total of 16 additional simulations (4 source groups × 4 simulation cases) were conducted over domain 4.

The temporal variations of source contributions during P1, P2, and P3 are shown in Figure 7, which gives an overview of source characteristics at the Tokyo site; this figure is based on the CMAQ base-case simulation. The first part of P1 was dominated by contributions from other sources but the latter part, when an increased SO$_4^{2-}$ concentration was revealed, was dominated by domestic sources of g01. Subsequently, P2 was dominated mainly by the contributions from other sources, and the contribution from g02 was the second largest factor. The contribution from g01 was small. During the latter part of P2, where there was an increase in SO$_4^{2-}$ concentration, the contribution from g01 was again observed. The period P3 was illustrated by declining contributions of other sources and g01, and the source contribution from g02 was negligible during P3. Throughout the entire period, source contributions from g03 and g04 were small because the g03 and g04 sources did not include SO$_2$ emissions.

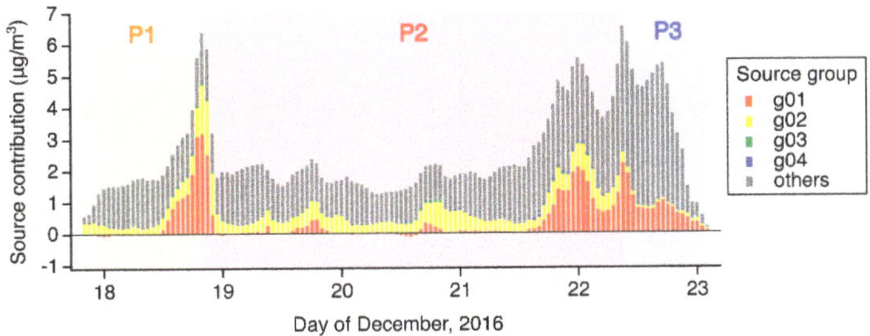

Figure 7. Temporal variation of source contributions of g01, g02, g03, g04, and others calculated for the CMAQ base-case simulation.

The spatial distributions of g01, g02, and other sources calculated in the CMAQ base-case simulation, shown in Figure 8, are based on the averages of data over P1, P2, and P3. During P1, the source contributions of g01 and g02 were limited over the Tokyo Bay area, and the source contributions of other sources, which were spread over the whole domain, were less than 2.0 μg/m^3 on average. These spatial distribution patterns for the source contributions were well understandable in terms of the simulated SO_4^{2-} concentrations shown in Figure 5. This clearly indicated that the high concentrations of SO_4^{2-}, which were greater than 2.0 μg/m^3 and limited to the coastline of Tokyo Bay, were due to the domestic contributions of g01 and g02. The impacts by domestic sources of g01 and g02 were spread over the Tokyo Bay area during P2, and the source contribution of g01 extended to the north during P3, whereas that of g02 was transported further into the northwestern area. The transportation of the source contribution of g02 in a northwestern direction may reflect a vertical emission distribution. Throughout P2 and P3, the source contributions of other sources were found over the whole domain. Therefore, it can be concluded that the simulated SO_4^{2-} concentrations over the Kanto region were generally dominated by the source contribution of others and the higher concentrations of SO_4^{2-} seen at the Tokyo observation site were attributed to the domestic sources of g01 and g02, which had limited impact around the Tokyo Bay area.

The similarities and differences of the source contributions conducted through the CMAQ base-case simulation and chemistry updates A, B, and KMT are summarized in Figure 9. The similarity of source contributions calculated by the CMAQ base-case simulation and the CMAQ KMT simulation are indicated for P1, P2, and P3. As revealed by the statistical analysis data listed in Table 3, this similarity was due to the similarity in model performances between the base-case simulation and KMT. Differences were found for chemistry updates A and B, and the magnitude of the differences was distinguishable during P1 compared with P2 and P3. Chemistry updates A, which was revised to enhance aqueous-phase oxidations, led to an increase in the source contribution of other sources. This result was also found in our previous study on the second phase of J-STREAM [4]. The aqueous-phase reactions considered were a refinement of the Fe- and Mn-catalyzed reactions and an additional NO_2 oxidation pathway; in this context, the concentrations of Fe, Mn, and NO_2 in Japan were much lower than those over Asia, thus the domestic source contributions were not enhanced. The approach taken in x was important for capturing the impact of transboundary air pollution.

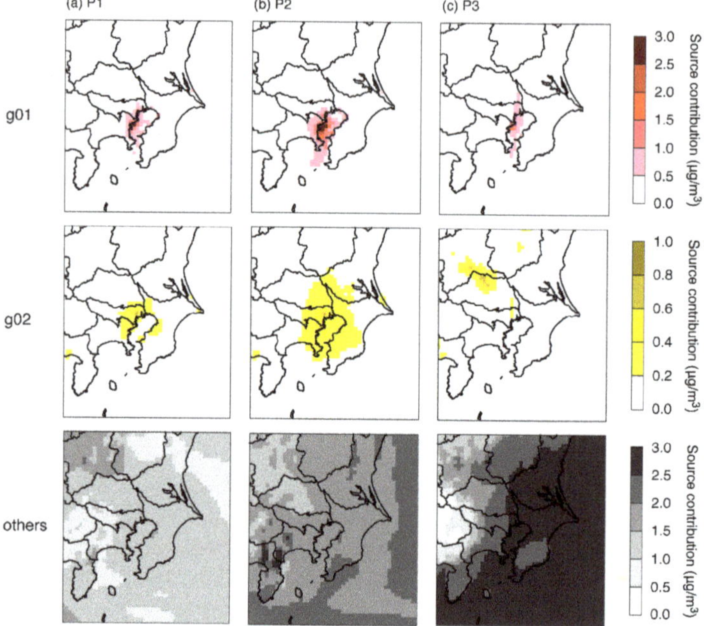

Figure 8. Spatial distributions of source contributions g01, g02, and other sources calculated by CMAQ base-case simulation, averaged during the analyzed periods defined as (**a**) P1, (**b**) P2, and (**c**) P3. Note that the color scale for g02 is different from that of g01 and other sources.

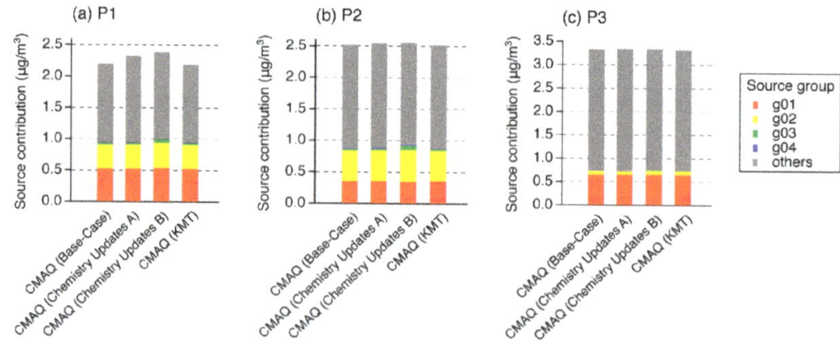

Figure 9. Summary of the source contributions calculated by the four CMAQ simulations averaged during (**a**) P1, (**b**) P2, and (**c**) P3.

Chemistry updates B, which included three SCI species, led to an increase in the source contributions of g03. For other simulations, there was negligible impact by g03; however, the source contribution of g03 was found in chemistry updates B, especially during P1. The source group g03 is a domestic fugitive VOC source and did not contain SO_2 emissions directly related to SO_4^{2-} production. This is because a change in the O_3 concentration. The source contributions of g01, g02, g03, and g04 on chemistry updates B during P1 are shown in Figure 10. The source contributions of g01 and g02 to the O_3 concentration were negative. The sources g01 and g02 contain abundant NOx emissions and urban areas generally correspond to VOC-sensitive regimes during the winter [40], hence reflecting the NOx disbenefit effect. The source contribution of g03 was related to an increase in the O_3 concentration, and this increment of the O_3 concentration was further connected to the increase in the three SCI species as incorporated in this modeling study, and which yielded an increase in SO_4^{2-} production. It was demonstrated that revision of the gas-phase SO_4^{2-} oxidation via SCI can be connected to the source sensitivity of VOC sources via a change in the O_3 concentration.

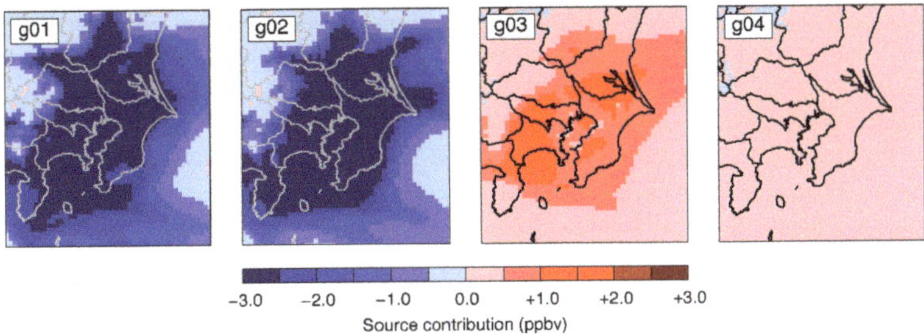

Figure 10. Source contributions of O_3 from four domestic sources of g01, g02, g03, and g04, calculated by chemistry updates B, averaged during P1.

4. Conclusions

The Japanese air quality model intercomparison study, J-STREAM, has shown that although SO_4^{2-} is generally well-captured by the models, concentrations of SO_4^{2-} were underestimated during winter. In previous studies, the modeled SO_4^{2-} concentrations were revised by focusing on the Fe- and Mn-catalyzed oxidation pathways and highlighting the importance of developing an emission inventory for trace metals over Asia in the first phase [3]. We further increased production of SO_4^{2-} by the addition of an aqueous-phase NO_2 oxidation pathway in the second phase [4]. To further improve the modeling performance of gas-phase oxidation, three SCI were incorporated in the third phase of J-STREAM and the KMT option available in the latest CMAQ model was examined. A difference between the KMT and the base-case simulation was found over the Sea of Okhotsk, and the absolute differences in SO_4^{2-} concentrations were less than -0.1 $\mu g/m^3$, giving essentially similar results with a test case for KMT over the U.S.A. Most previous studies have treated SCI in bulk; however, in this study, three SCI were treated separately with each SCI and incrementally tested and where the dependency of the rate constant of SCI1 with H_2O was also examined by performing sensitivity simulations with a high and low rate constant. It was found that only when the lower rate constant for the SCI1 + H_2O reaction was used that the production of SO_4^{2-} was increased by SCI1, and the importance of the value of the rate constant of SCI1 with H_2O for the Asian region was highlighted. This finding was consistent with a previous study conducted for the U.S.A. It was further demonstrated in the present study that the key role of SCI3 is to increase SO_4^{2-} production because this reaction is independent of the rate constant of SCI1 with H_2O. It was established that the explicit treatment of each SCI is required to enable clarification of the role of SCI on SO_4^{2-} production.

In addition to the investigation of model performance, the third phase of J-STREAM included an intercomparison study on source sensitivities. Four major domestic sources (transportation, stationary combustion, fugitive VOC, and agricultural NH_3) were investigated as source groups. The source sensitivities were estimated based on the traditional sensitivity simulation approach whereby a 20% emission reduction was calculated, the result of which was subtracted from the base-case simulation. It was clarified that the winter haze episode at the Tokyo site was generally dominated by emission sources from outside Japan, and the haze was enhanced by the domestic emission sources of transportation and fuel combustion. The estimations of source contributions were nearly the same between the base-case CMAQ simulation and KMT. With the chemistry updates involving the aqueous-phase Fe- and Mn-catalyzed oxidation reactions and NO_2 oxidation, it was found that these revisions led to an increase in transboundary impacts. In the case of the chemistry updates with the inclusion of SCI, it was shown that the change to fugitive VOC emissions could impact SO_4^{2-} concentrations by influencing O_3 which in turn influences SCI.

As a result of conducting the first, second and third phases of J-STREAM, we have successfully demonstrated a means to enhance simulated SO_4^{2-} production during the winter when underestimations of SO_4^{2-} concentrations have been problematic. Given recent drastic reductions in SO_2 emissions from China, further declines in SO_4^{2-} can be expected, and reactive nitrogen will continue to play an important role in this process due to the abundance of freely available NH_3. Because of the difficulty of producing reliable simulation models for reactive nitrogen species because of their semi-volatile nature, it is first necessary to establish accurate simulations for SO_4^{2-}. The means to enhance SO_4^{2-} production has been demonstrated for a single winter haze episode, and further tests on other haze episodes should be performed. Furthermore, the incorporation of SCI in this study suggests sensitivity to fugitive VOC sources that do not include direct SO_2 emissions but can change O_3 concentrations, and this effect should be tested in other seasons.

Author Contributions: S.I. coded the additional gas- and aqueous-phase reactions into the CMAQ model, conducted the model simulations and the comparison of models with observations, and wrote the manuscript; K.Y. is the sub-leader of the model intercomparison and prepared the meteorological inputs and initial and boundary conditions; H.H. is the sub-leader of the inorganic aerosol measurements and conducted the ACSA observations at Mukoujima, Tokyo; S.C. is the leader of the J-STREAM project and prepared the emission inputs and discussed the model intercomparison results.

Funding: This research was funded by the Environment Research and Technology Development Fund (5-1601) of the Environmental Restoration and Conservation Agency.

Acknowledgments: This research was supported by the Environment Research and Technology Development Fund (5-1601) of the Environmental Restoration and Conservation Agency.

Conflicts of Interest: The authors declare no conflict of interest.

References

1. Chatani, S.; Yamaji, K.; Sakurai, T.; Itahashi, S.; Shimadera, H.; Kitayama, K.; Hayami, H. Overview of model inter-comparison in Japan's study for reference air quality modeling (J-STREAM). *Atmosphere* **2018**, *9*, 19. [CrossRef]

2. Chatani, S.; Okumura, M.; Shimadera, H.; Yamaji, K.; Kitayama, K.; Matsunaga, S. Effects of a detailed vegetation database on simulated meteorological fields, biogenic VOC emissions, and ambient pollutant concentrations over Japan. *Atmosphere* **2018**, *9*, 179. [CrossRef]

3. Itahashi, S.; Yamaji, K.; Chatani, S.; Hayami, H. Refinement of modeled aqueous-phase sulfate production via the Fe- and Mn-catalyzed oxidation pathway. *Atmosphere* **2018**, *9*, 132. [CrossRef]

4. Itahashi, S.; Yamaji, K.; Chatani, S.; Hisatsune, K.; Saito, S.; Hayami, H. Model performance differences in sulfate aerosol in winter over Japan based on regional chemical transport models of CMAQ and CAMx. *Atmosphere* **2018**, *9*, 488. [CrossRef]

5. Kitayama, K.; Morino, Y.; Yamaji, K.; Chatani, S. Uncertainties in O_3 concentrations simulated by CMAQ over Japan using four chemical mechanisms. *Atmos. Environ.* **2019**, *198*, 448–462. [CrossRef]

6. Fu, X.; Wang, S.; Zhao, B.; Xing, J.; Cheng, Z.; Liu, H.; Hao, J. Emission inventory of primary pollutants and chemical speciation in 2010 for the Yangtze River Delta region, China. *Atmos. Environ.* **2013**, *70*, 39–50. [CrossRef]

7. Zheng, B.; Tong, D.; Li, M.; Liu, F.; Hong, C.; Geng, G.; Li, H.; Li, X.; Peng, L.; Qi, J.; et al. Trends in China's anthropogenic emissions since 2010 as the consequence of clean air actions. *Atmos. Chem. Phys.* **2018**, *18*, 14095–14111. [CrossRef]

8. Skamarock, W.C.; Klemp, J.B.; Dudhia, J.; Gill, D.O.; Barker, D.M.; Duda, M.G.; Huang, X.Y.; Wang, W.; Power, J.G. *A Description of the Advanced Research WRF Version 3; NCAR/TN-475+STR*; National Center for Atmospheric Research: Boulder, CO, USA, 2008.

9. National Centers for Environmental Prediction/National Weather Service/NOAA/U.S. Department of Commerce. NCEP GDAS/FNL 0.25 Degree Global Tropospheric Analyses and Forecast Grids, Research Data Archive at the National Center for Atmospheric Research, Computational and Information Systems Laboratory, Boulder, Colo. 2015. Available online: https://rda.ucar.edu/datasets/ds083.3/ (accessed on 7 May 2019).

10. Group for High Resolution Sea Surface Temperature (GHRSST). Available online: https://www.ghrsst.org (accessed on 7 July 2019).

11. Iacono, M.J.; Delamere, J.S.; Mlawer, E.J.; Shephard, M.W.; Clough, S.A.; Collins, W.D. Radiative forcing by long-lived greenhouse gases: Calculations with the AER radiative transfer models. *J. Geophys. Res.* **2008**, *113*, D13103. [CrossRef]

12. Morrison, H.; Thompson, G.; Tatarskii, V. Impacts of cloud microphysics on the development of trailing stratiform precipitation in a simulated squall line: Comparison of one- and two-moment schemes. *Mon. Weather Rev.* **2009**, *137*, 991–1007. [CrossRef]

13. Grell, G.A.; Devenyi, D. A generalized approach to parameterizing convection combining ensemble and data assimilation techniques. *Geophys. Res. Lett.* **2002**, *29*. [CrossRef]

14. US EPA Office of Research and Development. *Community Multiscale Air Quality (CMAQ) Model Version 5.2*; US EPA Office of Research and Development: Washington, DC, USA, 2017. [CrossRef]

15. Carter, W.P.L. Development of the SAPRC-07 chemical mechanism. *Atmos. Environ.* **2010**, *44*, 5336–5345. [CrossRef]

16. CMAQ v5.0 Sulfur Chemistry. Available online: https://www.airqualitymodeling.org/index.php/CMAQv5.0_Sulfur_Chemistry (accessed on 10 May 2019).

17. Seinfeld, J.H.; Pandis, S.N. *Atmospheric Chemistry and Physics—From Air Pollution to Climate Change*, 2nd ed.; John Wiley & Sons: New York, NY, USA, 2006.

18. Akimoto, H. *Atmospheric Reaction Chemistry*; Springer: New York, NY, USA, 2016.

19. Hatakeyama, S.; Akimoto, H. Reactions of Criegee intermediates in the gas phase. *Res. Chem. Intermed.* **1994**, *20*, 503–524. [CrossRef]

20. Welz, O.; Savee, J.D.; Osborn, D.L.; Vasu, S.S.; Percival, C.J.; Shallcross, D.E.; Taatjes, C.A. Direct kinetic measurements of Criegee Intermediate (CH_2OO) formed by reaction of CH_2I with O_2. *Science* **2012**, *335*, 204–207. [CrossRef]

21. Stone, D.; Blitz, M.; Daubney, L.; Howes, N.U.M.; Seakins, P. Kinetics of CH_2OO reactions with SO_2, NO_2, NO, H_2O and CH_3CHO as a function of pressure. *Phys. Chem. Chem. Phys.* **2014**, *16*, 1139. [CrossRef]

22. Tadayon, S.V.; Foreman, E.S.; Murray, C. Kinetics of the reactions between the Criegee intermediate CH_2OO and alcohols. *J. Phys. Chem. A* **2018**, *122*, 258–268. [CrossRef]

23. Taatjes, C.A.; Welz, O.; Eskola, A.J.; Savee, J.D.; Scheer, A.M.; Shallcross, D.E.; Rotavera, B.; Lee, E.P.F.; Dyke, J.M.; Mok, D.K.W.; et al. Direct measurements of conformer-dependent reactivity of the Criegee intermediate CH_3CHOO. *Science* **2013**, *340*, 177–180. [CrossRef]

24. Huang, H.-L.; Chao, W.; Lin, J.M. Kinetics of a Criegee intermediate that would survice high humidity and may oxidize atmospheric SO_2. *Proc. Natl. Acad. Sci. USA* **2015**, *112*, 10857–10862. [CrossRef]

25. Sarwar, G.; Fahey, K.; Kwok, R.; Gilliam, R.C.; Roselle, S.J.; Mathur, R.; Xue, J.; Yu, J.; Carter, W.P.L. Potential impacts of two SO_2 oxidation pathways on regional sulfate concentrations: aqueous-phase oxidation by NO_2 and gas-phase oxidation by Stabilized Criegee Intermediates. *Atmos. Environ.* **2013**, *68*, 186–197. [CrossRef]

26. Li, J.; Ying, Q.; Yi, B.; Yang, P. Role of stabilized Criegee Intermediates in the formation of atmospheric sulfate in eastern United States. *Atmos. Environ.* **2013**, *79*, 442–447. [CrossRef]

27. CMAQ v5.1 Aqueous Chemistry. Available online: https://www.airqualitymodeling.org/index.php/CMAQv5.0_Sulfur_Chemistry (accessed on 28 June 2018).

28. Fahey, K.M.; Carlton, A.G.; Pye, H.O.T.; Baek, J.; Hutzell, W.T.; Stanier, C.O.; Baker, K.R.; Appel, K.W.; Jaoui, M.; Offenberg, J.H. A framework for expanding aqueous chemistry in the Community Multiscale Air Quality (CMAQ) model version 5.1. *Geosci. Model Dev.* **2017**, *10*, 1587–1605. [CrossRef]

29. Damian, V.; Sandu, A.; Damian, M.; Potra, F.; Carmichael, G.R. The kinetic preprocessor KPP—A software environment for solving chemical kinetics. *Comput. Chem. Eng.* **2002**, *26*, 1567–1579. [CrossRef]

30. Itahashi, S.; Uno, I.; Kim, S.-T. Source contributions of sulfate aerosol over East Asia estimated by CMAQ-DDM. *Environ. Sci. Technol.* **2012**, *46*, 6733–6741. [CrossRef]

31. Itahashi, S.; Hayami, H.; Yumimoto, K.; Uno, I. Chinese province-scale source apportionments for sulfate aerosol in 2005 evaluated by the tagged tracer method. *Environ. Pollut.* **2017**, *220*, 1366–1375. [CrossRef]

32. Itahashi, S.; Hatakeyama, S.; Shimada, K.; Tatsuta, S.; Taniguchi, Y.; Chan, C.K.; Kim, Y.-P.; Lin, N.-H.; Takami, A. Model estimation of sulfate aerosol source collected at Cape Hedo during an intensive campaign in October-November, 2015. *Aerosol Air Qual. Res.* **2017**, *17*, 3079–3090. [CrossRef]

33. Itahashi, S. Toward synchronous evaluation of source apportionments for atmospheric concentration and deposition of sulfate aerosol over East Asia. *J. Geophys. Res. Atmos.* **2018**, *123*, 2927–2953. [CrossRef]

34. Itahashi, S.; Hatakeyama, S.; Shimada, K.; Takami, A. Sources of high sulfate aerosol concentration observed at Cape Hedo in spring 2012. *Aerosol Air Qual. Res.* **2019**, *19*, 587–600. [CrossRef]
35. Itahashi, S.; Uno, I.; Osada, K.; Kamiguchi, Y.; Yamamoto, S.; Tamura, K.; Wang, Z.; Kurosaki, Y.; Kanaya, Y. Nitrate transboundary heavy pollution over East Asia in winter. *Atmos. Chem. Phys.* **2017**, *17*, 3823–3843. [CrossRef]
36. EANET. Technical Manual for Wet Deposition Monitoring in East Asia. Available online: http://www.eanet. asia/product/manual/techwet.pdf (accessed on 3 September 2018).
37. Emery, C.; Liu, Z.; Russell, A.G.; Odman, M.T.; Yarwood, G.; Kumar, N. Recommendations on statistics and benchmarks to assess photochemical model performance. *J. Air Waste Manag. Assoc.* **2017**, *67*, 582–598. [CrossRef]
38. Boylan, J.W.; Russell, A.G. PM and light extinction model performance metrics, goals, and criteria for three-dimensional air quality models. *Atmos. Environ.* **2006**, *40*, 4946–4959. [CrossRef]
39. Fiore, A.M.; Dentener, F.J.; Wild, O.; Cuvelier, C.; Schultz, M.G.; Hess, P.; Textor, C.; Schulz, M.; Doherty, R.M.; Horowitz, L.W.; et al. Multimodel estimates of intercontinental source-receptor relationships for ozone pollution. *J. Geophys. Res.* **2009**, *114*, D04301. [CrossRef]
40. Itahashi, S.; Uno, I.; Kim, S.-T. Seasonal source contributions of tropospheric ozone over East Asia based on CMAQ-HDDM. *Atmos. Environ.* **2013**, *70*, 204–217. [CrossRef]

Article

Analysis of Pollutant Dispersion in a Realistic Urban Street Canyon Using Coupled CFD and Chemical Reaction Modeling

Franchesca G. Gonzalez Olivardia *, Qi Zhang, Tomohito Matsuo, Hikari Shimadera and Akira Kondo

Graduate School of Engineering, Osaka University, 2-1 Yamadaoka, Suita, Osaka 565-0871, Japan
* Correspondence: franchesca@ea.see.eng.osaka-u.ac.jp; Tel.: +81-6-6879-7668

Received: 31 July 2019; Accepted: 19 August 2019; Published: 21 August 2019

Abstract: Studies in actual urban settings that integrate chemical reaction modeling, radiation, and particular emissions are mandatory to evaluate the effects of traffic-related air pollution on street canyons. In this paper, airflow patterns and reactive pollutant behavior for over 24 h, in a realistic urban canyon in Osaka City, Japan, was conducted using a computational fluid dynamics (CFD) model coupled with a chemical reaction model (CBM-IV). The boundary conditions for the CFD model were obtained from mesoscale meteorological and air quality models. Inherent street canyon processes, such as ground and wall radiation, were evaluated using a surface energy budget model of the ground and a building envelope model, respectively. The CFD-coupled chemical reaction model surpassed the mesoscale models in describing the NO, NO_2, and O_3 transport process, representing pollutants concentrations more accurately within the street canyon since the latter cannot capture the local phenomena because of coarse grid resolution. This work showed that the concentration of pollutants in the urban canyon is heavily reliant on roadside emissions and airflow patterns, which, in turn, is strongly affected by the heterogeneity of the urban layout. The CFD-coupled chemical reaction model characterized better the complex three-dimensional site and hour-dependent dispersion of contaminants within an urban canyon.

Keywords: CFD; chemical reaction model; urban canyon; radiation; mesoscale models; reactive pollutants

1. Introduction

Elevated levels of air pollution generated by the road traffic have been associated with severe illness such as the development of asthma in adults, impaired lung growth in children, and an increased risk of chronic obstructive pulmonary disease [1]. Consequently, air quality in urban canyons needs to be improved to protect pedestrians, cyclists, and drivers from exposure to contaminants such as particulate matter (PM), volatile organic compounds (VOCs), carbon monoxide (CO), nitrogen oxides (NO_x), and sulfur dioxide (SO_2) that are primarily emitted from motor vehicles [2]. Among these contaminants, NO_x and VOCs represent the most significant concern for air pollution in street canyons because they are released at very short distances from the receptors. Moreover, under the presence of sunlight, fast reactions occur between them leading to the formation of particulate matter and tropospheric ozone (O_3) which are very dangerous secondary pollutants [3].

Various tools have been employed to measure the levels of contaminants inside the street canyons [4], such as, field observations [5–7], wind tunnels measurements [8,9], and several semi-empirical models including the ADMS-Urban dispersion model [10], CALINE4 [11], and the Urban Street Model [12].

The technique utilized for this study was computational fluid dynamics (CFD) modeling because it helps to reduce the time required to optimize a physical model and is less expensive in comparison with wind tunnels. Moreover, the simulation of dispersion and boundary conditions with CFD are better studied [13]. Additionally, recent advances in computer technology make CFD a powerful tool for air pollution modeling [14,15]; using CFD, the description of air flow patterns at the city-block scale with high spatial-resolution can be achieved [16], and important simulations such as the influence of building geometry on wind flow and heat convection can be carried out [17,18].

In microscale environments, many factors such as vehicle emissions, street canyon configuration, surface heating, and chemical reactions are dynamically involved in the dispersion of pollutants. A number of studies have tried to model these factors using CFD. To evaluate surface heating, Kim and Baik [19] systematically characterized two-dimensional flow regimes with the presence of street bottom heating, although failed to take heating of buildings into account. Xie et al. [20] considered the impact of heating of three idealized sunlit wall configurations on pollutant dispersion in a two-dimensional computational domain perpendicular to the street canyon wind direction; later, Kim and Baik [21] extended this work by examining the effects of street-bottom and building-roof heating on flow fields in a three-dimensional ideal urban canyon setting. This work highlighted the importance of including street-bottom and building roof heating in three-dimensional flow models. Bottillo et al. [22] remarked on the effect of solar radiation in an urban street canyon with three-dimensional flow field under different ambient wind conditions, confirming the impact of the velocity on the heat transfer process.

Meanwhile, studies about the dispersion of reactive pollutants inside street canyon were conducted by Baker et al. [23] using a simple NO–NO_2–O_3 photochemistry model to evaluate the isothermal dispersal and chemistry in and above the street canyons with an aspect ratio of 1. Baik et al. [24] extended this work by showing, with a simple street canyon model and an aspect ratio of 1, the importance of the building surface radiation effects on the wind flow pattern in the urban areas when modeling reactive pollutant dispersion. In the following years, Kwak and Baik [25] calculated the photochemical reaction of pollutants in and above an idealized street canyon using a two-dimensional CFD model coupled for the first time with a Carbon Bond Mechanism IV (CBM-IV) [26]. Similarly, Kwak and Baik [27] used an idealistic two-dimensional model with urban surface, radiation and chemical processes to focus on the diurnal variation of NO_x and O_3 exchange. This study also employs a CFD model coupled with a CBM-IV chemical model to represent reactive pollutants inside the street canyons.

Despite significant effort, there is still a lack of research in non-idealized three-dimensional environments which restricts the ability to simulate pollutant concentrations at the street level in complex urban areas. One of the main reasons for this is the specification of boundary conditions for realistic urban areas. Some authors have overcome this drawback by coupling mesoscale models with CFD simulations, for example, Baik et al. [28] used a CFD model coupled to a mesoscale meteorological model to examine urban flow and pollutant dispersion in a densely built-up area of Seoul. Tewari et al. [29] reported the benefits of coupling a microscale transport and dispersion model with a mesoscale numerical weather prediction model; significant improvements were observed when wind fields were produced by downscaling a meteorological model output as the initial and boundary CFD conditions. Recently, Kwak et al. [30] developed an integrated three-dimensional model that coupled CFD with mesoscale meteorological and chemistry-transport models to simulate the air pollution from 09:00 to 18:00 local time in Seoul. In this work, the Weather Research and Forecasting model (WRF) [31] and the Community Multiscale Air Quality model (CMAQ) [32] are employed as mesoscale meteorological and air quality models to obtain boundary conditions for CFD.

This study is an ambitious attempt to evaluate the concentration of reactive pollutants in a realistic urban street canyon domain over 24 h by integrating a CFD model coupled with a chemical reaction model (CBM-IV), a radiation model, and boundary conditions from WRF-CMAQ. The first objective is to couple CFD with the chemical reaction model (CBM-IV) to describe urban dynamics with high resolution. The second is to validate the CFD-coupled chemical reaction model by comparing the numerical simulation results of a realistic urban street canyon against the observed data collected from monitoring stations situated near the roadside. The final objective is to assess the roadside pollutant dispersion in realistic city blocks with the CFD-coupled chemical reaction model.

This paper is structured as follows: in Section 2, the numerical CFD model coupled with the CBM-IV model, the surface energy budget model of the ground, and the building envelope model are presented. All the simulation settings required for the modeling of a realistic urban canyon are given in Section 3. In Section 4, the results of the CFD-coupled chemical reaction model simulations are validated against the observed data and CMAQ-simulated data. Also in the same section, the wind flow patterns and distribution of NO, NO_2, and O_3 at 08:00, 12:00, 16:00, and 20:00 JST are discussed in detail. In the last section, the conclusions are presented.

2. CFD-Coupled Chemical Reaction Model

In this study, the dispersion of reactive pollutants in urban canyons is described by the equations of continuity (Equation (1)), momentum (Equations (2) and (3)), and scalar transport coupled with the CBM-IV chemical model (Equation (4)). Air is modeled as an incompressible viscous flow, and the buoyancy forces are treated with the Boussinesq approximation.

Continuity equation:

$$\frac{\partial U_j}{\partial x_j} = 0, \tag{1}$$

Momentum equation:

$$\frac{\partial U_i}{\partial t} + U_j \frac{\partial U_i}{\partial x_j} = -\frac{1}{\rho_0}\frac{\partial P^*}{\partial x_i} + \delta_{i3} g \frac{T^*}{T_0} + v \frac{\partial^2 U_i}{\partial x_j \partial x_j} - \frac{\partial}{\partial x_j}\left(\overline{u_i u_j}\right), \tag{2}$$

$$-\overline{u_i u_j} = K_m\left(\frac{\partial U_i}{\partial x_j} + \frac{\partial U_j}{\partial x_i}\right) - \frac{2}{3}\delta_{ij}k, \tag{3}$$

where x_i represents the ith cartesian coordinates, U_i is the ith mean velocity component (m/s), t is the time (s), ρ is the air density (kg/m^3), P^* is the pressure (Pa) derivation from its reference value, δ_{ij} is the Kronecker delta, g is the gravitational acceleration (m/s^2), T^* is the temperature derivation from its reference value ($=T - T_0$) (K), T_0 is the reference temperature (K), v is the kinematic viscosity of air (m^2/s), K_m is the eddy diffusivity of momentum (m^2/s), and k is the turbulent kinetic energy (m^2/s^2).

Scalar transport equation coupled with CBM-IV model:

$$\frac{\partial C_i}{\partial t} + U_j \frac{\partial C_i}{\partial x_j} = D\frac{\partial^2 C_i}{\partial x_j \partial x_j} + \frac{\partial}{\partial x_j}\left(K_c \frac{\partial C_i}{\partial x_j}\right) + \left[\frac{\partial C_i}{\partial t}\right]_{Chem}, \tag{4}$$

where C_i represents the mean concentration of ith species in the air (ppm), D is the molecular diffusivity of the species (m^2/s), and K_c is the eddy diffusivity of the species (m^2/s). The last term of the scalar transport equation represents the net chemical production rate of the species calculated by the CBM-IV model.

The turbulence model standard k-ε is employed in this study (Equations (5) and (6)).

Turbulent kinetic energy equation:

$$\frac{\partial k}{\partial t} + U_j \frac{\partial k}{\partial x_j} = -\overline{u_i u_j}\frac{\partial U_i}{\partial x_j} + \frac{\delta_{3j} g}{T_0}\overline{T' u_j} + \frac{\partial}{\partial x_j}\left(\frac{K_m}{\sigma_k}\frac{\partial k}{\partial x_j}\right) - \varepsilon, \tag{5}$$

where σ_k is an empirical constant, ε is the dissipation rate (m^2/s^3).

Turbulent dissipation rate of energy equation:

$$\frac{\partial \varepsilon}{\partial t} + U_j \frac{\partial \varepsilon}{\partial x_j} = -C_{\varepsilon 1} \frac{\varepsilon}{k} \overline{u_i u_j} \frac{\partial U_i}{\partial x_j} + -C_{\varepsilon 1} \frac{\varepsilon}{k} \frac{\delta_{3j}}{T_0} \overline{T' u_j} + \frac{\partial}{\partial x_j}\left(\frac{K_m}{\sigma_\varepsilon} \frac{\partial \varepsilon}{\partial x_j} \right) - C_{\varepsilon 2} \frac{\varepsilon^2}{k}, \tag{6}$$

where $C_{\varepsilon 1}$ and $C_{\varepsilon 2}$ represent empirical constants.

Moreover, the energy equation (Equation (7)) is used to evaluate the influence of the radiation on the flow pattern and dispersion of pollutants.

Energy equation:

$$\frac{\partial T}{\partial t} + U_j \frac{\partial T}{\partial x_j} = \alpha \frac{\partial^2 T}{\partial x_j \partial x_j} + \frac{\partial}{\partial x_j}\left(\alpha_h \frac{\partial T}{\partial x_j} \right) + \frac{Q}{C_p \rho}, \tag{7}$$

where T represents the mean temperature (K), α is the dispersion by molecular motion (m^2/s), α_h is the dispersion by eddies (turbulence) (m^2/s) and Q represents the source term of heat (W/m^2), C_p is the specific heat at a constant pressure (J/kg/K), and ρ is the density of the air (kg/m^3). As part of this work, considerations about the influence of shortwave radiation (direct solar radiation) and longwave radiation (diffuse solar radiation) are taken into account by a surface energy budget model of the ground [33] shown as follows:

$$\left(1 - \alpha_g\right) S \downarrow -\sigma T_i^4 + f_{i,sky} R \downarrow + \sigma \sum f_{i,j} T_j^4 - H = Q_G, \tag{8}$$

where α_g is the ground albedo, $S \downarrow$ is shortwave radiation (W/m^2), σ is Stefan–Boltzmann constant (W/m^2/K^4), T_i is the temperature of an element in the ground surface (K), $f_{i,sky}$ is the sky view factor for element i, $R \downarrow$ is downward longwave radiation (W/m^2), $f_{i,j}$ is the view factor of each surface element j from element i, T_j is the temperature of surface j (K), H is the sensible heat flux (W/m^2), and Q_G is the heat flux into the ground (W/m^2). In Equation (8), the emissivity for longwave radiation is omitted because it is assumed to be 1 for all surfaces.

Similarly, a slightly modified building envelope model [34] accounted for the radiation received in the air from the walls and roof surfaces of buildings (Equation (9)).

$$(1 - \alpha_{bs}) S \downarrow -\sigma T_{bso}^4 + f_{i,sky} R \downarrow + \sigma \sum f_{bso,j} T_j^4 + \rho_0 C_p C_H u (T_{bso} - T_{ao}) = Q_w, \tag{9}$$

where α_{bs} is the building surface albedo, T_{bso} is the temperature of the outer building wall (K), c_p is the specific heat at a constant pressure (J/kg/K), C_H is the bulk heat transfer coefficient (W/m^{2}K), T_{ao} (K) is the temperature of air above T_{bso}, and Q_w is heat flux from the building into the air (W/m^2).

The finite volume method is used to discretize all the equations. The first-order Euler method is utilized as the time discretization scheme and the power law scheme as the discretization scheme of the convection and diffusion terms in the governing equations. Moreover, the semi-implicit method for pressure-linked equations (SIMPLE) algorithm was employed as the pressure–velocity coupling algorithm [35]. The solar radiation and view factor are calculated following the method of Ikejima et al. [33].

3. Simulation Setup

The three-dimensional CFD domain was set up in Umeda-Shinmichi (34.70° N, 135.50° E), Osaka City, Japan (Figure 1a), where there is a typical urban street canyon layout with five roads flanked by twenty-two buildings (Figure 1b). The information about the length, width, and height of every building was obtained by the results of the survey on land use conducted by Osaka City in 2005. The height of the array of buildings was from 12 to 48 m.

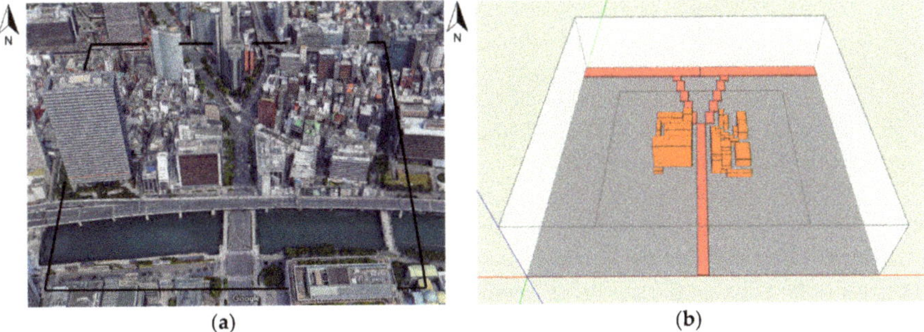

(a)   (b)

Figure 1. Calculated domain (**a**) Umeda-Shinmichi, Google Maps 2018 (34.70° N, 135.50° E); (**b**) computational fluid dynamics (CFD) representation of the three-dimensional city block.

The size of the calculation domain was 600 m × 600 m × 150 m, and the analysis area was defined as 200 m × 200 m × 150 m, red square in Figure 2a. The hexahedral mesh was 93 × 93 × 51 in the x-, y- and z-directions, respectively (Figure 2), and a grid interval of 3.33 m in x- and y-direction was set.

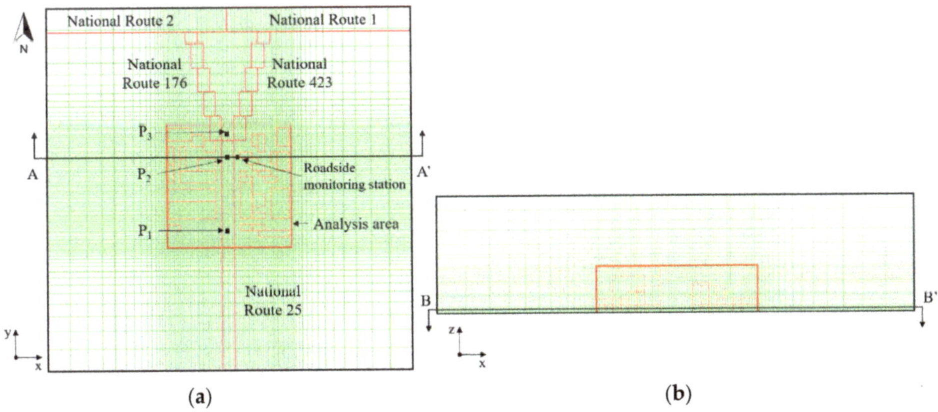

(a)   (b)

Figure 2. Mesh view of the CFD calculation domain and analysis area. (**a**) the x–y view shows the reference points P_1 located at y = 230 m, point P_2 and cross-section A-A′ located at y = 350 m, and point P_3 located at y = 390 m; (**b**) x–z view shows the cross-section B-B′ located at z = 1.5 m.

The boundary conditions of the CFD domain for temperature, wind speed and wind direction, and those for air pollutant concentrations were obtained from results of meteorological simulations by WRF v3.7 (Supplementary Material Figure S1) and air quality simulations by CMAQ v5.1 (Supplementary Material Figure S2), respectively. Supplementary Material Figure S3 and Table S1 correspondingly show modeling domains and configurations for the WRF-CMAQ simulation. The simulation was conducted over seven modeling domains from domain 1 (D1), covering East Asia with 64-km grids, to domain 7 (D7), covering Osaka Prefecture with 1-km grids, for a period from 24 June to 31 August 2010. The vertical domain consisted of 30 layers with the middle heights of the first, second, and third layers being 28 m, 92 m, and 190 m, respectively. The physics parameterizations and input data for WRF and the chemical mechanisms, including the Carbon Bond mechanism developed in 2005 (CB05) [36], the next generation model of CBM-IV, for CMAQ were the same as those used by Shimadera et al. [37]. Emission data for CMAQ were produced from the same datasets as those used by Uranishi et al. [38], including the Japan Auto-Oil Program Emission Inventory-Data Base for vehicles (JEI-DB) in the year 2010 developed by the Japan Petroleum Energy Center [39]. As shown in Supplementary Material

Figure S3, there are substantial NO$_x$ emissions from vehicles, industrial areas, and large point sources in Osaka City.

23 August 2010 was chosen as the calculation date for the 24-h unsteady state CFD analysis because of the very clear and calm conditions so that photochemical reactions might play an essential role in air quality on the day. The WRF and CMAQ simulations in D7 on the day were compared with observed data by the Japan Meteorological Agency (JMA) at the Osaka Meteorological Observatory (34.68° N, 135.52° E) (Supplementary Material Figure S3) and by the Ministry of the Environment of Japan (MOE) at Kokusetsu-Osaka station (34.68° N, 135.54° E) (Supplementary Material Figure S3). Figure 3 shows diurnal variations of the observed and WRF-simulated ground-level air temperature, wind speed and direction at the Osaka Meteorological Observatory, which is located 2 km southeast of Umeda-Shinmichi, on 23 August. The WRF-simulated air temperature agreed reasonably well with the observed data, showing a Root Mean Square Error (RMSE) value of 0.48 °C. The model also approximately captured the temporal variation of wind speed with a slight overestimation, indicated with a RMSE of 0.71 m/s. Both in the observation and WRF simulation, the wind direction mainly ranged from southwest to west, except at 08:00 and 09:00 JST (UTC+9) in the observation under calm conditions. For the case of the wind direction, the RMSE was 43 degrees.

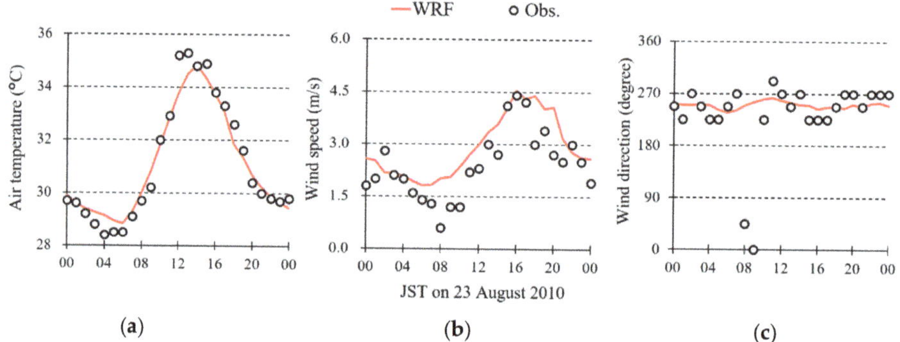

Figure 3. Diurnal variations of Weather Research and Forecasting model (WRF)-simulated and observed (**a**) air temperature; (**b**) wind speed; (**c**) wind direction at the Osaka Meteorological Observatory on 23 August 2010.

Figure 4 shows the diurnal variation of observed and CMAQ-simulated ambient concentrations of NO, NO$_2$, and O$_3$ at the Kokusetsu-Osaka station, which is located 4 km east-southeast of Umeda-Shinmichi. CMAQ approximately captured the variations and magnitudes of ambient concentrations of these air pollutants, except for NO$_2$ which was underestimated during the daytime. The RMSE of the NO, NO$_2$, and O$_3$ were 1.39 ppb, 7.03 ppb, and 11.43 ppb, respectively. These results indicate that WRF-CMAQ successfully produced meteorological and concentration fields around the study area for the CFD boundary conditions.

Because of the coarser vertical resolution of the WRF-CMAQ model compared with CFD, Monin–Obukhov Similarity Theory (MOST) was employed to determine the boundary vertical distribution of air temperature and wind components. Under MOST, the WRF-CMAQ data at heights 28 m and 92 m of Umeda-Shinmichi in D7 were used with a roughness length of 0.1 m. Moreover, the concentration under 28 m was the same value as the data at 28 m, and over 28 m the values were interpolated linearly. Because CB05 used in CMAQ has more species of VOCs than CBM-IV used in CFD, VOC species in CMAQ output were lumped into those in CBM-IV. Once this data was obtained, the vertical distribution from WRF-CMAQ at every hour and the values interpolated linearly in between each hour, the boundary conditions at transient state were set for the CFD model.

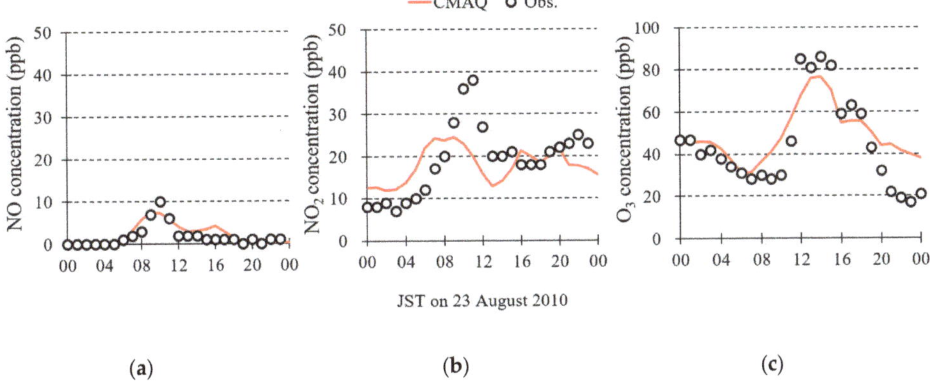

Figure 4. Diurnal variations of Community Multiscale Air Quality model (CMAQ)-simulated and observed (**a**) NO, (**b**) NO$_2$, and (**c**) O$_3$ concentrations at the Kokusetsu-Osaka station for monitoring ambient air pollution on 23 August 2010.

The emission rate from vehicles used as CMAQ input data was derived from JEI-DB with a horizontal resolution of 1 km × 1 km. The total emission rate from automobiles in the CFD domain was estimated from the JEI-DB data by multiplying the area ratio of the CFD domain to a grid of JEI-DB (0.36). Then, the total emission was allocated into the five national routes shown in Figure 2 (National Route 1, National Route 2, National Route 176, National Route 423, National Route 25) by considering the traffic volume of each road provided by the Japan Ministry of Land, Infrastructure, Transport and Tourism [40]. Figure 5 shows the diurnal variation of the NO$_x$ (NO + NO$_2$) emission rate for 24 h for every national route. The average NO$_2$/NO$_x$ emission ratio for the indicated area (0.36) was 0.1799. The information about the NO$_x$ emission rate for every national route was used as the boundary condition for the emission rate and updated every hour in the CFD-coupled chemical reaction model. As discussed above, models of 24 h are mandatory since rush hours represent the peak emission rates.

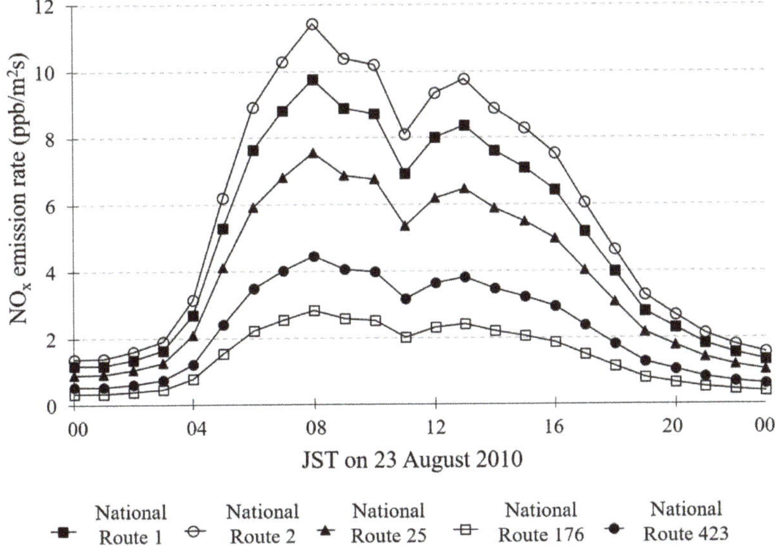

Figure 5. Diurnal variation of NO$_x$ (NO + NO$_2$) emissions rates on 23 August 2010.

4. Results and Discussion

The dispersion of pollutants inside urban street canyons is a phenomenon heavily reliant on the building geometry and roadside emissions. Therefore, it is quite challenging to capture the local aspects in mesoscale simulations softwares, such as WRF or CMAQ, because low model resolution cannot be well characterized [41]. In Figure 6, the validation of the CFD model coupled chemical reaction model (CBM-IV) against CMAQ-simulated and the roadside monitoring station data by the Atmospheric Environmental Regional Observation System (AEROS) at Umeda-Shinmichi station (34.69° N, 135.50° E) about the diurnal variations of NO and NO_2 on 23 August 2010, is shown. For the case of O_3 in Figure 6c, the comparison is only made between CMAQ and CFD simulations since the roadside monitoring station did not record the data for Ozone.

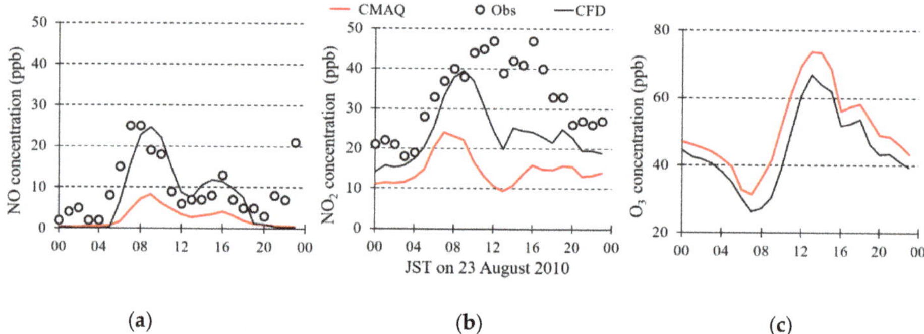

Figure 6. Diurnal variations in CMAQ-simulated, observed, and CFD-simulated (**a**) NO; (**b**) NO_2; (**c**) O_3 concentrations at the Umeda-Shinmichi station for monitoring roadside air pollution on 23 August 2010.

The RMSE in the case of the CFD-simulated results for NO and NO_2 were 6.35 ppb and 11.44 ppb, however, for the CMAQ-simulated results the values of the RMSE were 9.08 ppb and 20.35 ppb, respectively. Compared with CMAQ-simulated results, CFD-coupled chemical reaction model is more accurate describing the NO and NO_2 concentrations. The NO levels from the CFD calculation behaved almost identical to the observed data. As it is shown, there are some underestimations in NO_2 concentration from 09:00 JST to 18:00 JST, most likely because of the underestimation of the daytime background concentration indicated in Figure 4. The O_3 concentrations output from CMAQ and CFD simulations show similar behavior, however the CFD simulations show consistently lower concentration. The results of this validation indicate how important it is to study particular emissions, the reactions in the modeling of air pollution, and the actual configuration of the urban street canyon using coupled CFD and chemical reaction modeling.

The behavior of NO, NO_2, and O_3 inside the analysis area, where the main road, National Route 25, runs perpendicular to the wind inflow with a width of 20 m, and sidewalks of 10 m at both sides was investigated (Figure 2). The emission line source was placed in the middle of National Route 25 with a width of 20 m and a height of 1 meter from the ground. Figure 7 shows the airflow pattern in the analysis area at 08:00, 12:00, 16:00, and 20:00 JST in cross-section B-B' (Figure 2b). The height of z = 1.5 m is the same height as the monitoring station at which the data was compared. The flow field in Figure 7 indicates a wind circulation restriction behind buildings typically present in street canyons. Throughout the day the inlet wind direction was sustained from west to east, and from afternoon (16:00 JST) until nighttime (20:00 JST) when the velocity reached maximum levels. Inside the street canyon (National Route 25), the wind flow moved mainly from south to north through the entire street canyon because of open space in the southwest part of the analysis area. Moreover, counterclockwise vortices developed on the road at 08:00, 12:00, 16:00, and 20:00 JST behind the buildings on the left side of the road, as result of the wind that speeded up and went through the channels between constructions. At noon, a shift in the wind flow in the superior side of the analysis area from west to east was observed, creating more vortices between buildings located downstream. Additionally, Figure 7 demonstrates the importance of three-dimensional analyses since some eddies are seen at buildings corners where shear stresses are common and responsible for some turbulence in the flow.

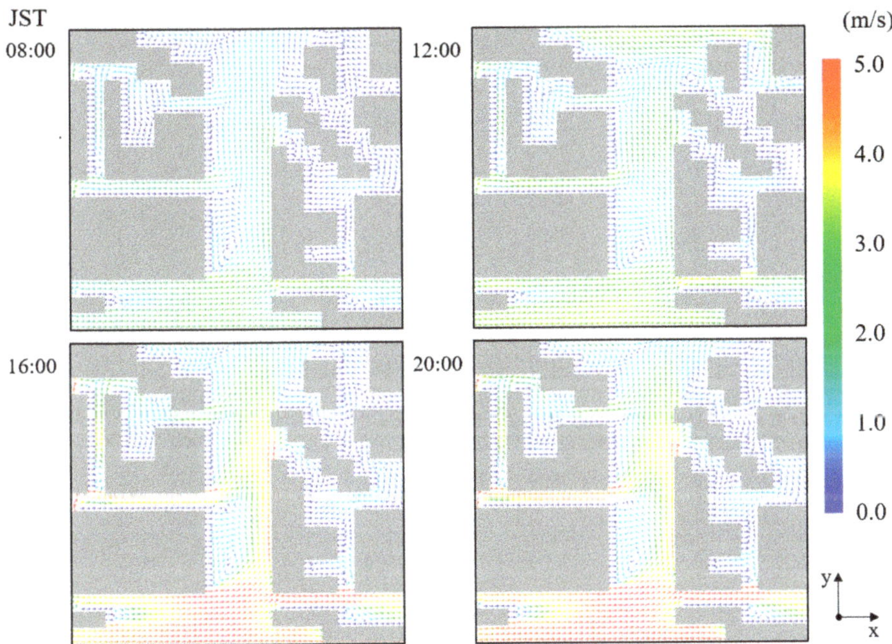

Figure 7. Airflow patterns at 08:00, 12:00, 16:00, and 20:00 JST, cross-section B-B' view (Figure 2b).

Figure 8 shows the spatial distribution of NO, NO_2, and O_3 concentrations at 08:00, 12:00, 16:00, and 20:00 JST in cross-section B-B' (Figure 2b). High levels of NO and NO_2 remained in the middle of the street canyon because building configurations do not allow their removal. During the day, the behavior of the contaminants changes depending on the emission rates and wind velocity. During the morning rush hours (07:00–09:00 JST), the NO_x concentration in the street canyon increases (Figure 5) and stays on the street from 08:00 JST until 12:00 JST, especially behind the buildings where vortices are observed (Figure 7). At 16:00 JST, the vehicle emission rate decreases, and the wind speed is higher, so the NO_2 concentrations reduced. However, at 20:00 JST, some slight concentration of NO_2 is still observed because of the titration reaction between NO and O_3 [42]. Throughout the day, it is observed in Figure 6, a different distribution between O_3 and NO_2 due to the proximity of the traffic-related emission (NO) that leads to NO_x titration. Besides, it is shown in Figure 8 that high levels of NO and NO_2 are located in the right superior quadrant at noon because of the shift of the wind direction observed in Figure 7.

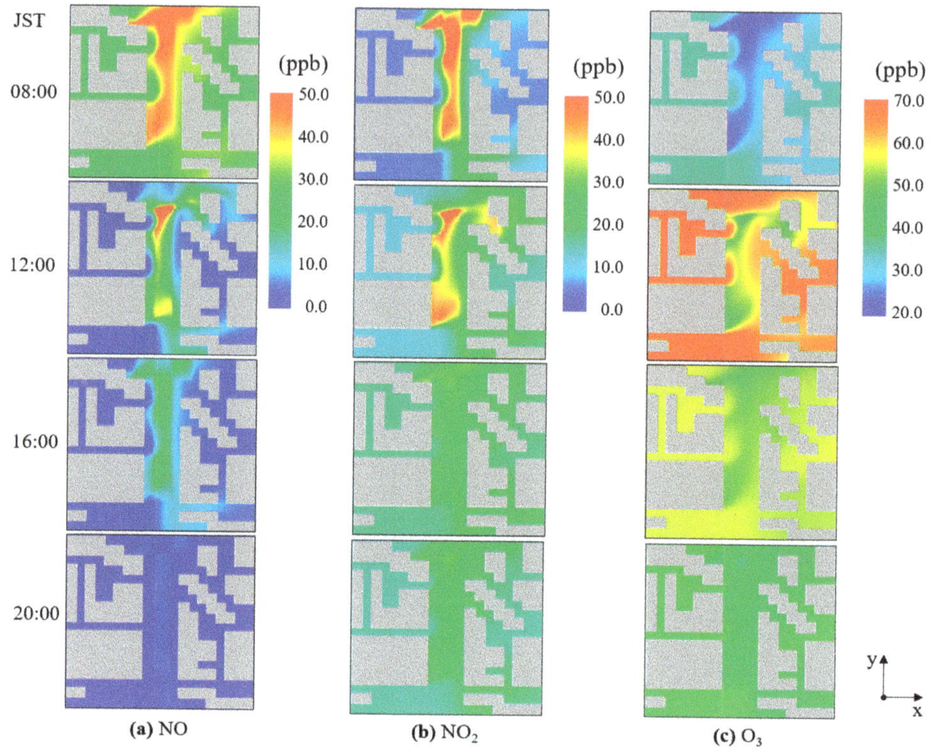

Figure 8. Spatial distribution of (**a**) NO, (**b**) NO$_2$, and (**c**) O$_3$ concentration at 08:00, 12:00, 16:00, and 20:00 JST, cross-section B-B' view (Figure 2b).

Figure 9 shows the airflow patterns at 08:00, 12:00, 16:00, and 20:00 JST in cross-section A-A' (Figure 2a). Weaker wind flow is seen on the leeward side of the buildings and stagnation of the flow is evident upwind which is characteristic of wind flow around the buildings [43]. When the wind flow encounters the buildings the flow is separated and the wind speed decreases, depending on the magnitude of the wind speed the scale of the vortices changes when moving in the positive z-direction. For this lateral view (cross-section A-A', Figure 2a), isolated roughness flow is observed, however, some perturbation is induced because of the proximity of the buildings in the background of the cross-section A-A'.

As mentioned before, adequate characterization of the dispersion and reaction of pollutants inside a real street urban environment requires the modeling of the effects of buildings and ground heating. The results of the analysis of the influence of the temperature on the analysis area, in specific Point 1, Point 2, and Point 3 (Figure 2a) are presented in the Supplementary Material Figure S6. The analysis of the temperature profiles showed no considerable difference under and above the heights of the buildings (~30 m). Thus, the influence of air temperature on the transport of contaminants for this urban street layout was considered minimal and almost negligible in comparison with the wind flow.

Figure 10 shows the spatial distribution of (a) NO, (b) NO$_2$, and (c) O$_3$ concentrations at 08:00, 12:00, 16:00, and 20:00 JST at cross-section A-A' (Figure 2a). Because of the vehicle exhaust located at a height of 1 meter in the middle of the analysis area, high NO and NO$_2$ concentrations remained near the surface level, mainly from 08:00 JST to 12:00 JST. The dispersion plume of NO, NO$_2$, and O$_3$ exhibits large vertical gradient from the ground to roof level of the building. As mentioned by Melkonyan and Kuttler [44], NO and NO$_2$ enhance ozone's dissociation and production, respectively. Thus, if the NO/NO$_2$ ratio decreases, ozone concentrations increase. In this study, all the National Routes studied showed high emission rates with a NO/NO$_2$ ratio of 4:1, suggesting that the NO$_x$ titration was the main reason for the dissociation of O$_3$. At 20:00 JST, even when emissions from vehicles had decreased, the NO$_2$ concentration in the street was still present in contrast with the concentration above roof level. The vertical exchange of air pollutants is, therefore, shown to be mainly influenced by the vehicle emissions and building features.

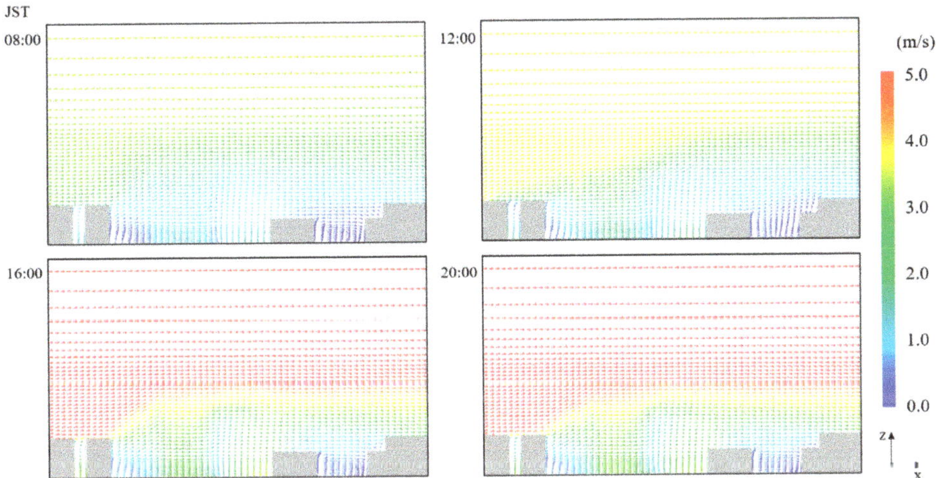

Figure 9. Airflow patterns at 08:00, 12:00, 16:00, and 20:00 JST, cross-section A-A' view (Figure 2a).

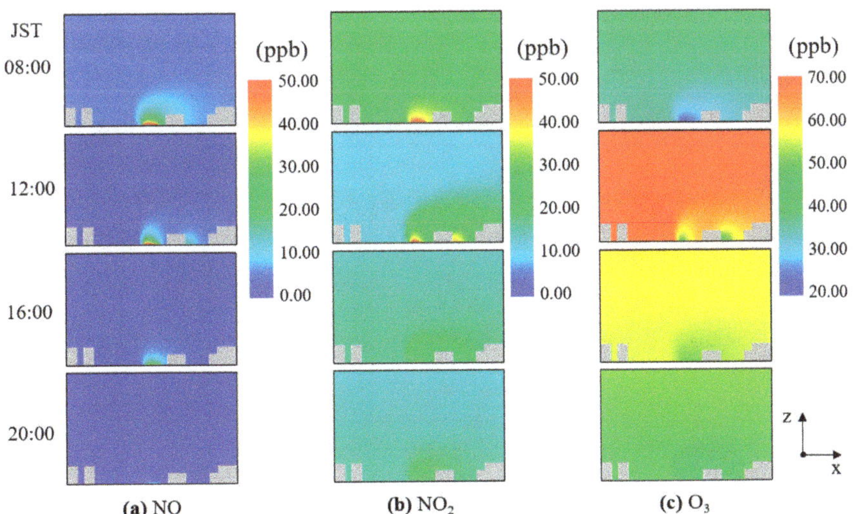

Figure 10. Spatial distribution of (a) NO, (b) NO$_2$, and (c) O$_3$ concentration at 08:00, 12:00, 16:00, and 20:00 JST, cross-section A-A' view (Figure 2a).

5. Conclusions

This paper investigated the behavior of reactive pollutants inside a realistic urban street canyon by coupling a CFD model with a chemical reaction model (CBM-IV). The complexity of the urban street canyon geometry was represented by the CFD model and the dynamical mechanism involved in the dispersion of the pollutants were incorporated using mesoscale and radiation models. The spatial distribution of the reactive pollutants, in particular, NO, NO_2, and O_3 were researched over a 24-h period on 23 August 2010. The dispersion of the contaminants was highly dependent on the reaction processes, boundary conditions, and emission rates all integrated at the same time within the urban street canyon. The production of NO_x or fading of O_3 were especially found in regions with low wind speed and high turbulence, and NO_x titration was noted to be of great importance. The O_3 behavior was directly affected by the chemical reactions near the roadside, where fresh NO was being emitted, and was consequently controlled by the NO_x distribution. The grid resolution of WRF-CMAQ appears to have a strong influence when representing the boundary conditions, and still, because of their limitations (1 km × 1 km minimum grid size) the particularities that accompany urban areas such as street, highways, unequal height of buildings, sidewalks cannot be well represented.

Finally, we can conclude that the prevailing wind flow mainly carried the air pollutants in the windward direction with small vortices recirculating pollutants inside the street canyon. This work is an intent to find better representations of boundary conditions and to step forward in the incorporation of radiation models and reactive models with the purpose of simulating urban-like environments, satisfactorily. Further efforts in this kind of research are necessary to reproduce the realities of the urban areas and the implications that it may have on the people who are exposed to these concentrations during the day.

Supplementary Materials: The following are available online at http://www.mdpi.com/2073-4433/10/9/479/s1, Figure S1: Boundary conditions for air temperature and wind speed at 08:00, 12:00, 16:00, and 20:00 JST. Figure S2: Boundary conditions for NO, NO_2, and O_3 concentrations at 08:00, 12:00, 16:00, and 20:00 JST. Figure S3: Spatial distributions of mean NO_x emission intensity in CMAQ modeling domains from D1 covering East Asia to D7 covering Osaka Prefecture, and locations of observation sites in Osaka City used for model validations. Figure S4: Diurnal variations of the CMAQ-simulated and observed NO_2/NO_x concentration ratio at the Kokusetsu-Osaka station for monitoring ambient air pollution on 23 August 2010. Figure S5: Diurnal variations of the CMAQ-simulated, observed and CFD-simulated NO_2/NO_x concentration ratio at the Umeda-Shinmichi station for monitoring roadside air pollution on 23 August 2010. Figure S6: Vertical air temperature profiles at 08:00, 12:00, 16:00, and 20:00 JST on National Route 25 for the points (**a**) P_1 located at y = 230 m; (**b**) P_2 located at y = 350 m; (**c**) P_3 located at y = 390 m. Table S1: WRF and CMAQ configurations.

Author Contributions: F.G.G.O. analyzed the data and also wrote the paper; Q.Z. performed the numerical simulations; T.M., H.S., and A.K. provided significant suggestions on data analyses. All the authors have read and approved the final manuscript.

Funding: This research received no external funding.

Conflicts of Interest: The authors declare no conflict of interest.

References

1. Laumbach, R.J.; Kipen, H.M. Respiratory health effects of air pollution: Update on biomass smoke and traffic pollution. *J. Allergy Clin. Immunol.* **2012**, *129*, 3–13. [CrossRef] [PubMed]
2. Faiz, A.; Weaver, C.S.; Walsh, M.P. *Air Pollution from Motor Vehicles: Standards and Technologies for Controlling Emissions*, 1st ed.; The World Bank: Washington, DC, USA, 1996.
3. Canadian Council of Ministers of the Environment. NOx/VOC Fact Sheets 1998. 2014. Available online: https://www.ccme.ca/en/resources/air/emissions.html/ (accessed on 15 March 2018).
4. Vardoulakis, S.; Fisher, B.E.A.; Pericleous, K.; Gonzalez-Flesca, N. Modelling air quality in street canyons: A review. *Atmos. Environ.* **2003**, *37*, 155–182. [CrossRef]
5. DePaul, F.T.; Sheih, C.M. Measurements of wind velocities in a street canyon. *Atmos. Environ.* **1986**, *20*, 455–459. [CrossRef]
6. Qin, Y.; Kot, S.C. Dispersion of vehicular emission in street canyons, Guangzhou city, South China (P.R.C.). *Atmos. Environ.* **1993**, *27*, 283–291. [CrossRef]
7. Vardoulakis, S.; Gonzalez-Flesca, N.; Fisher, B.E.A. Assessment of traffic-related air pollution in two street canyons in Paris: Implications for exposure studies. *Atmos. Environ.* **2002**, *36*, 1025–1039. [CrossRef]
8. Ahmad, K.; Khare, M.; Chaudhry, K.K. Wind tunnel simulation studies on dispersion at urban street canyons and intersections—A review. *J. Wind Eng. Indus. Aerodyn.* **2005**, *93*, 697–717. [CrossRef]

9. Cui, P.-Y.; Li, Z.; Tao, W.-Q. Wind-tunnel measurements for thermal effects on the air flow and pollutant dispersion through different scale urban scales. *Build. Environ.* **2016**, *97*, 137–151. [CrossRef]

10. Righi, S.; Lucialli, P.; Pollini, E. Statistical and diagnostic evaluation of the ADMS-Urban model compared with an urban air quality monitoring network. *Atmos. Environ.* **2009**, *43*, 3850–3857. [CrossRef]

11. Kenty, K.L.; Poor, N.D.; Kronmiller, K.G.; McClenny, W.; King, C.; Atkeson, T.; Campbell, S.W. Application of CALINE4 to roadside NO/NO$_2$ transformations. *Atmos. Environ.* **2007**, *41*, 4270–4280. [CrossRef]

12. Taseiko, V.; Mikhailuta, S.V.; Pitt, A.; Lezhenin, A.A.; Zakharov, Y.V. Air pollution dispersion within urban street canyons. *Atmos. Environ.* **2009**, *43*, 245–252. [CrossRef]

13. Leitl, M.; Meroney, R.N. Car exhaust dispersion in a street canyon. Numerical critique of a wind tunnel experiment. *J. Wind Eng. Ind. Aerodyn.* **1997**, *67–68*, 293–304. [CrossRef]

14. Kim, J.-J.; Baik, J.-J. A numerical study of thermal effects on flow and pollutant dispersion in urban street canyons. *J. Appl. Meteorol.* **1999**, *38*, 1249–1261. [CrossRef]

15. Baik, J.-J.; Kim, J.-J.; Fernando, H.J.S. A CFD model for simulating urban flow and dispersion. *J. Appl. Meteorol.* **2003**, *42*, 1636–1648. [CrossRef]

16. Solazzo, E.; Vardoulakis, S.; Cai, X. A novel methodology for interpreting air quality measurements from urban streets using CFD modelling. *Atmos. Environ.* **2011**, *45*, 5230–5239. [CrossRef]

17. Dayang, W.; Yun, Z.; Qingxiang, L.; Xuesong, D.; Yong, Z. Numerical study on wind-pressure characteristics of a high rise building in group of buildings. In Proceedings of the Seventh International Colloquium on Bluff Body Aerodynamics and Applications, Shanghai, China, 2–6 September 2012; pp. 521–530.

18. de Lieto Vollaro, A.; Galli, G.; Vallati, A. CFD analysis of convective heat transfer coefficient on external surfaces of buildings. *Sustainability* **2015**, *7*, 9088–9099. [CrossRef]

19. Kim, J.-J.; Baik, J.-J. Urban street-canyon flows with bottom heating. *Atmos. Environ.* **2001**, *35*, 3395–3404. [CrossRef]

20. Xie, X.; Huang, Z.; Wang, J.; Xie, Z. The impact of solar radiation and street layout on pollutant dispersion in street canyon. *Build. Environ.* **2005**, *40*, 201–212. [CrossRef]

21. Kim, J.-J.; Baik, J.-J. Effects of street-bottom and buildings-roof heating on flow in three-dimensional street canyons. *Adv. Atmos. Sci.* **2010**, *27*, 513–527. [CrossRef]

22. Bottillo, S.; de Lieto Vollaro, A.; Galli, G.; Vallati, A. CFD modeling of the impact of solar radiation in a tridimensional urban canyon at different wind conditions. *Sol. Energy* **2014**, *102*, 212–222. [CrossRef]

23. Baker, J.; Walker, H.L.; Cai, X. A study of the dispersion and transport of reactive pollutants in and above street canyons- a large eddy simulation. *Atmos. Environ.* **2004**, *38*, 6883–6892. [CrossRef]

24. Baik, J.-J.; Kang, Y.-S.; Kim, J.-J. Modeling reactive pollutant dispersion in an urban street canyon. *Atmos. Environ.* **2007**, *41*, 934–949. [CrossRef]

25. Kwak, K.-H.; Baik, J.-J. A CFD modeling study of the impacts of NOx and VOC emissions on reactive pollutant dispersion in and above a street canyon. *Atmos. Environ.* **2012**, *46*, 71–80. [CrossRef]

26. Gery, M.W.; Whitten, G.Z.; Killus, J.P.; Dodge, M.C. A photochemical kinetics mechanism for urban and regional scale computer modeling. *J. Geophys. Res.* **1989**, *94*, 12925–12956. [CrossRef]

27. Kwak, K.-H.; Baik, J.-J. Diurnal variation of NOx and ozone exchange between a street canyon and the overlying air. *Atmos. Environ.* **2014**, *86*, 120–128. [CrossRef]

28. Baik, J.-J.; Park, S.-B.; Kim, J.-J. Urban flow and dispersion simulation using a CFD model coupled to a mesoscale model. *J. Appl. Meteorol. Clim.* **2009**, *48*, 1667–1681. [CrossRef]

29. Tewari, M.; Kusaka, H.; Chen, F.; Coirier, W.J.; Kim, S.; Wyszogrodzki, A.A.; Warner, T.T. Impact of coupling a microscale computational fluids dynamics model with a mesoscale model on urban scale contaminant transport and dispersion. *Atmos. Res.* **2010**, *96*, 656–664. [CrossRef]

30. Kwak, K.-H.; Baik, J.-J.; Ryu, Y.-H.; Lee, S.-H. Urban air quality simulation in a high-rise building area using a CFD model coupled with mesoscale meteorological and chemistry-transport models. *Atmos. Environ.* **2015**, *100*, 167–177. [CrossRef]

31. Skamarock, W.C.; Klemp, J.B. A time-split nonhydrostatic atmospheric model for weather research and forecasting applications. *J. Comput. Phys.* **2008**, *227*, 3465–3485. [CrossRef]

32. Byun, D.; Schere, K.L. Review of the governing equations, computational algorithms, and other components of the Models-3 Community Multiscale Air Quality (CMAQ) modeling system. *Appl. Mech. Rev.* **2006**, *59*, 51–77. [CrossRef]

33. Ikejima, K.; Kondo, A.; Kaga, A. The 24-h unsteady analysis of air flow and temperature in a real city by high-speed radiation calculation method. *Build. Environ.* **2011**, *46*, 1632–1638. [CrossRef]

34. Cortes, A.; Murashita, Y.; Matsuo, T.; Kondo, A.; Shimadera, H.; Inoue, Y. Numerical evaluation of the effect of photovoltaic cell installation on urban thermal environment. *Sustain. Cities Soc.* **2015**, *19*, 250–258. [CrossRef]

35. Patankar, S.V. *Numerical Heat Transfer and Fluid Flow*, 1st ed.; Taylor & Francis: New York, NY, USA, 1980; pp. 126–129.

36. Yarwood, G.; Rao, S.; Yocke, M.; Whitten, G. *Updates to the Carbon Bond Chemical Mechanism: CB05, Final Report to the US EPA, RT-0400675*; YOCKE AND COMPANY: Novato, CA, USA, 2005.

37. Shimadera, H.; Kojima, T.; Kondo, A. Evaluation of Air Quality Model Performance for Simulating Long-Range Transport and Local Pollution of PM2.5 in Japan. *Adv. Meteorol.* **2016**, 1–13. [CrossRef]

38. Uranishi, K.; Ikemori, F.; Shimadera, H.; Kondo, A.; Sugata, S. Impact of field biomass burning on local pollution and long-range transport of PM2.5 in Northeast Asia. *Environ. Pollut.* **2019**, *244*, 414–422. [CrossRef]

39. Japan Petroleum Energy Center. *Technical report of the Japan Auto-Oil Program: Emission inventory of road transport in Japan*; JPEC-2011AQ-02-06; Japan Petroleum Energy Center (JPEC): Tokyo, Japan, 2012; p. 136. (In Japanese)

40. The Japanese Ministry of Land, Infrastructure, Transport and Tourism. National Road Traffic Volume Census. Available online: http://www.mlit.go.jp/road/census/h22-1/index.html/ (accessed on 27 June 2017). (In Japanese)

41. Martilli, A. Current research and future challenges in urban mesoscale modelling. *Int. J. Clim.* **2007**, *27*, 1909–1918. [CrossRef]

42. Sillman, S. The relation between ozone, NOx and hydrocarbons in urban and polluted rural environments. *Atmos. Environ.* **1999**, *33*, 1821–1845. [CrossRef]

43. Tominaga, Y.; Sato, Y.; Sadohara, S. CFD simulations of the effect of evaporative cooling from water bodies in a micro-scale urban environment: Validation and application studies. *Sustain. Cities Soc.* **2015**, *19*, 259–270. [CrossRef]

44. Melkonyan, A.; Kuttler, W. Long-term analysis of NO, NO_2 and O_3 concentrations in North Rhine-Westphalia, Germany. *Atmos. Environ.* **2012**, *60*, 316–326. [CrossRef]

Article

Determination of the Area Affected by Agricultural Burning

Daniel F. Prato [1] and Jose I. Huertas [2,*]

1   Centro Latinoamericano de Innovación en Logística (CLI), Bogotá 111071, Colombia; dprato@logyca.org
2   School of Engineering and Science, Energy and Climate Change Research Group,
     Tecnologico de Monterrey 64849, Mexico
*   Correspondence: jhuertas@tec.mx; Tel.: +52-1-722-100-9417

Received: 4 April 2019; Accepted: 27 May 2019; Published: 5 June 2019

Abstract: Agricultural burning is still a common practice around the world. It is associated with the high emission of air pollutants, including short-term climate change forcing pollutants such as black carbon and $PM_{2.5}$. The legal requirements to start any regulatory actions to control them is the identification of its area of influence. However, this task is challenging from the experimental and modeling point of view, since it is a short-term event with a moving area source of pollutants. In this work, we assessed this agricultural burning influence-area using the US Environmental authorities recommended air dispersion model (AERMOD). We considered different sizes and geometries of burning areas located on flat terrains, and several crops burning under the worst-case scenario of meteorological conditions. The influence area was determined as the largest area where the short-term concentrations of pollutants (1 h or one day) exceed the local air quality standards. We found that this area is a band around the burning area whose size increases with the burning rate but not with its size. Finally, we suggested alternatives of public policy to regulate this activity, which is based on limiting the burning-rate in the way that no existing households remain inside the resulting influence-area. However, this policy should be understood as a transition towards a policy that forbids agricultural burning.

Keywords: open burning; biomass burning; sugarcane crops; environmental assessment; air quality modeling

---

1. Introduction

Agricultural burning is the controlled incineration of biomass before and after harvesting. It is a common practice worldwide to harvest and to control and eliminate fungi and pests, reduce the erosion and maintain the soil quality for future crops at the lowest cost [1]. Despite the technological progress, currently, ~60% of the harvesting processes worldwide take place manually, which leads to the biomass burning over large areas of cultivated land (>1 ha per burning event) [2]. This practice has been widely studied for the case of wildfires, which has several implications on climate, atmospheric composition and air quality [3]. Presently, around 8600 Tg/year of biomass are burned globally, from which ~2000 Tg/year are related to agricultural crops [4]. Table 1 lists the main crops of interest for the environmental authorities. They correspond to those crops carried out extensively where open burning is a common practice. Now, the agricultural burning is mainly associated with industrial sugarcane crops, in countries such as Brazil, Colombia, Guatemala, India, Mexico and Costa Rica. During the harvesting period, biomass burning produces fine and ultrafine particles (particles with aerodynamic diameter $d < 30$ μm, and $d < 100$ nm, respectively) [2,5]. It also contributes to the emissions of carbon dioxide (CO_2), carbon monoxide (CO), methane (CH_4), hydrocarbons (HC), nitrogen oxides (NO_x), and other toxic compounds, such as Polycyclic aromatic hydrocarbons (PAHS) and Volatile organic compounds (VOCs), which are ozone precursors [6]. Agricultural burning could lead to short-term

(~1-day) episodes of air pollution, due to the capability of the emitted pollutants of being transported over a large spatial scale [7] and to their contribution to the formation of secondary pollutants such as ozone [8,9]. As a result, agricultural burning can cause adverse health effects on the people living nearby the burning areas [10,11].

Table 1. Emission factors and loading factors for different crops [12].

Crops	Emission Factors ($E^*_{i,j}$)						Loading Factors (L_j)
	PM	PM$_{2.5}$	CO	NO	Methane	Non-Methane Organic Compounds	
	g/Mg	kg/Mg	kg/Mg	kg/Mg	kg/Mg	kg/Mg	Mg/hectare
Unspecified	11	12.5	58	1.3	2.7	9	4.5
Asparagus	20	-	75	-	10	33	3.4
Barley	11	-	78	-	2.2	7.5	3.8
Corn	7	-	54	-	2	6	9.4
Cotton	4	-	88	-	0.7	2.5	3.8
Grasses	8	-	50	-	2.2	7.5	-
Pineapple	4	-	56	-	1	3	-
Rice	4	12.95	41	-	1.2	4	6.7
Safflower	9	-	72	-	3	10	2.9
Sorghum	9	-	38	-	1	3.5	6.5
Sugarcane	**2.3–3.5**	-	**30–41**	-	**0.6–2**	**2–6**	**8–46**
Alfalfa	23	-	53	-	4.2	14	1.8
Bean (red)	22	-	93	-	5.5	18	5.6
Hay (wild)	16	-	70	-	2.5	8.5	2.2
Oats	22	-	68	-	4	13	3.6
Pea	16	-	74	-	4.5	15	5.6
Wheat	11	4.71	64	-	2	6.5	4.3

* PM, particulate matter with aerodynamic diameter d < 30 μm; PM$_{2.5}$, particulate matter with aerodynamic diameter d < 2.5 μm.

Prado et al. [13] reported that, during the harvesting period in Mendonça, Brazil, the concentration of particulate matter registered in the atmosphere of urban areas, near to sugarcane fields, was almost 2.5-times higher than the World Health Organization air quality recommendation for short-term human exposure (24 h) [14]. Wagner et al. [15] measured ambient particle concentrations and particle type downwind, upwind and at several distances from agricultural burns in Imperial Valley, California. They reported significantly high PM$_{10}$ and PM$_{2.5}$ concentrations at locations less than 3.2 km from the nearest burning. Mugica et al. [16] estimated sugarcane-burning emissions in Mexican municipalities, and reported exceedances on the PM$_{2.5}$ Mexican emission standards by at least 5.4 times, with an average of 86 ± 22 μg m^{-3}. Their measurements were used to adjust the parameters of their Gaussian dispersion model, with which they studied 25 additional burning episodes. They observed concentrations up to 1000 μg/m^3 in urban areas when the wind blew towards those urban areas during the burning episodes.

Biomass burning is poorly regulated worldwide. Environmental authorities require the identification of the agricultural burning influence area as a legal requirement to start any regulatory actions to control this activity [17]. The influence area is defined as the largest area where the concentration of any pollutant exceeds the local National Atmospheric Air Quality Standards (NAAQS). The extent of the influence area depends on multiple factors including the size and geometry of the burning area, the local meteorological conditions and the pollutant considered. The influence area can be determined experimentally or by using any well-accepted dispersion model.

The experimental determination of the agricultural burning influence area possesses technical challenges due to the need for a large number of simultaneous measurements required for each variable affecting the dispersion phenomenon. Carney et al. [18] proposed a methodology to estimate the influence area as a cone whose orientation is aligned to the predominant wind direction. Aiming to advance this work, Hiscox et al. [17] measured the size and dispersion of smoke plumes during four sugarcane burning events during pre- and post-harvesting periods in Louisiana, USA, using a scanning, elastic-backscatter LiDAR (Laser Imaging Detection and Ranging). Their results show that particle concentration exceeded the NAAQS at distances of up to 300 m from the source, and that the vertical extension of the plume was about 2 km. They also found that wind speed and atmospheric stability conditions could make the plumes to travel distances greater than 45 km. Based on these studies, the USEPA developed guidelines that limit the meteorological conditions under which land cultivators of this region can burn [18]. The conclusions obtained with these experimental works are valid for the characteristics of the particular region, type of crops and meteorological conditions under which researchers conducted their experiments. The lack of generality of these conclusions limits the possibility of using them for an eventual policy to control agricultural burning in other regions.

Alternatively, air dispersion models can be used for estimating the size of the influence area under varying scenarios of meteorological conditions, crops types and area sizes. We propose the use of AERMOD for this purpose. It is a steady-state Gaussian dispersion model developed by the American Meteorological Society and the USEPA (The United States Environmental Protection Agency) for regulatory purposes. Gaussian dispersion models assume that pollutant concentration, downwind the source, follows a normal distribution in the horizontal and vertical direction. The main challenges of using this model for the study of the environmental impact of agricultural burning are:

- AERMOD requires that the burning area be considered as a fixed source of pollutant when, it actually is a moving area source.
- AERMOD requires input data for the mass emission of pollutants. They are estimated via emission factors. However, the determination of emission factors for open burning is challenging due to its diffusive nature. Usually, they are obtained from laboratory and field studies [16,19] and, as expected, there are large differences in the emission factors reported by researchers [12–20].

The use of AERMOD for the determination of the agricultural burning influence areas has not been extensively employed. In this work, we systematically used AERMOD to study the dispersion of the pollutants emitted from short-term agricultural burning events, under varying conditions of emission rates, meteorological conditions, sizes and geometries of the burning areas. Then, based on the obtained results of pollutant concentration downwind: (i) we identified the size of the generated influence area; and (ii) we proposed alternatives of public policy to control this activity.

2. Methodology

The dispersion of the atmospheric pollutants generated by the burning of several agricultural crops was simulated using AERMOD over a region with general characteristics (flat terrain) and under multiple meteorological conditions. We observed the size of the influence area generated on each case for different sizes and geometries of the burning areas. Aiming to facilitate the analysis, we identified the crop and the pollutant that produced the largest mass emission per unit of cultivated area. Next, we describe the regions studied, the meteorology considered, the estimation of the emission rates and the methodology used to determine the influence area.

2.1. Study Region

The most frequent cases of agricultural burning occur in areas located on relatively flat terrains enclosed by rectangular polygons, in the surroundings of urban centers [21]. As shown below, the determination of the influence area of agricultural burning occurring in areas delimited by non-regular polygons can be analyzed as a combination of multiple squared areas. Therefore, as a base case, we used a burning area of 1 ha. To evaluate the sensitivity of the model to the size of the burning area, we varied it from 1 m^2 to 20 ha and considered different area orientations. The determination of the influence area generated by agricultural burning in mountainous terrain requires special simulations for each region considered. Furthermore, traditional dispersion models present problems to estimate accurately the concentration of pollutants under those conditions. Therefore, they are out of the scope of the present work.

A grid of discrete receptors was defined within the computational domain with a resolution of 100 m over an area of 10 km × 10 km, as shown in Figure 1. Although the results do not depend on the location of the area under consideration, for illustrative purposes, we used the Valle del Cauca region, located in southwest Colombia, which is one of the most important areas of sugarcane cultivation. Colombia is the seventh producer of sugarcane worldwide, and where approximately 16,425 ha are burned per year, and where on average 1.5–2 ha are burned per burning event [19].

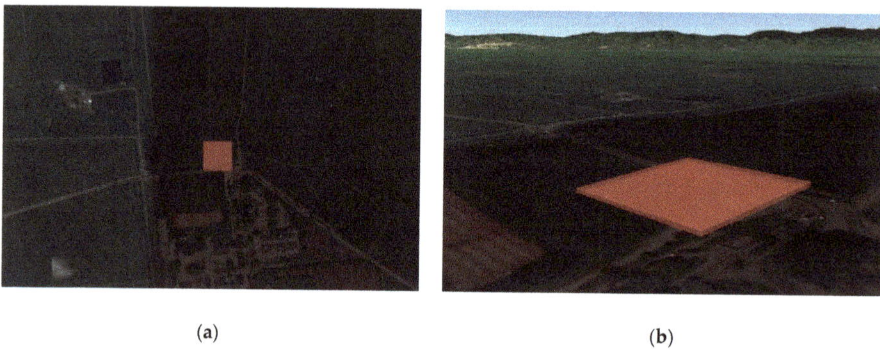

(a)   (b)

Figure 1. (**a**) Top; and (**b**) perspective view of the sugarcane area selected in this work to illustrate the estimation of the influence area generated by agricultural burning. The red square represents a burning area of 1 ha.

2.2. Air Dispersion Model

In this work, we used AERMOD to study the dispersion of the pollutants produced inside the burning area. As stated above, this model is recommended by the USEPA when their results are planned to be used for regulatory purposes. It is a steady-state model that assumes that pollutants concentration downwind the area source follows a Gaussian distribution in the vertical and horizontal direction of the plume, according to Equation (1).

$$C = \frac{Q}{u} \cdot \frac{f}{\sigma_y \sqrt{2\pi}} \cdot \frac{g}{\sigma_z \sqrt{2\pi}} \tag{1}$$

where C is the pollutant concentration (g/m^3); Q is the pollutant emission rate (g/s); u is the horizontal wind speed along the plume centerline (m/s); H is the height of emission plume centerline above ground level (m); σ_z and σ_y are the vertical and horizontal standard deviation of the emission distribution (m), respectively; and f and g are the vertical and horizontal dispersion parameters, respectively.

Gaussian models, and specifically AERMOD, do not allow, on the first instance, to model sources that change their position over time. As an approximation, we assumed that the entire emission source area burns simultaneously, but at a rate such that the emission rate of pollutants (g/s) remains constant. In Section 3, we will show that this assumption is acceptable for this study because the size of the influence area is independent of the size of the burning area being considered.

2.3. Meteorology

High wind speeds contribute to pollutant dispersion but it could also extend the size of the influence area. Low wind speeds generate high concentration of pollutants near the source. The fact that dispersion phenomena is highly dependent of the meteorological conditions hampers the process of generalizing the results obtained. To overcome this difficulty, we studied the dispersion of pollutants generated by agricultural burning under very diverse meteorological conditions. Datasets with 1–5 years of 1-h meteorological conditions from the USA and Colombia were used, since they represent scenarios with extreme weather characteristics during the different seasons of the year. Table 2 presents the list of meteorological data used in this study and Appendix A. shows their respective wind roses. Only meteorological datasets with 100% of data availability were used in work.

Table 2. Datasets of 1-h meteorological data used to study the dispersion of the pollutants generated during agricultural burnings.

ID	Meteorology	Country	Year
1	San Diego	USA	2009–2014
2	Minnesota	USA	2009–2013
3	Texas	USA	1990
4	Michigan	USA	1990
5	Alaska	USA	1990
6	Zavala	USA	2008–2012
7	Pico	USA	2008–2012
8	Descanso	Colombia	2009
9	Cerro largo	Colombia	2009
10	Rubiales	Colombia	2013
11	Los Angeles	USA	2012–2016

2.4. Estimation of the Pollutants' Mass Emission Rates

The mass emission rate $E_{i,j}$ (kg/s) of pollutant i emitted by a given crop burning j, was estimated through Equation (2), where L_j (kg/m^2) is the amount of biomass that is typically produced by crop j per unit of cultivated area, and S_j (m^2/s) is the burning rate [22].

$$E_{i,j} = E^*_{i,j}\, L_j\, S_j \tag{2}$$

In this equation, $E^*_{i,j}$ (kg/kg) is the emission factor and it describes the amount of pollutant i typically emitted per unit mass of crop j. Multiple studies have been conducted to determine the emission factors associated with agricultural burning under controlled conditions [1] and by field measurements [23]. Table 1 shows $E^*_{i,j}$ and L_j for several crops. It shows that sugarcane has the largest loading factor. For this crop, Table 3 presents the emission factors reported by different authors, among which, large variations are observed. In this study, we adopted the emission factors reported by the USEPA [22].

The burning rate (S_j in m^2/s) depends on multiple factors, including wind speed and crop moisture content. Given the difficulty of finding values reported in the literature for this variable, a constant value was assumed as a first approximation. For the case of sugarcane, we consulted companies in the sugarcane industry, and they reported an approximate value of 1 ha/day. However, it depends on the length and the number of lines used as starting flame fronts.

2.5. Determination of the Influence Area

For a given set of meteorological conditions (temperature, humidity, solar radiation, wind speed and direction), AERMOD estimates the pollutant concentration at every receptor located nearby the burning area. The process is repeated every hour as meteorological data are reported in this format. Given the short-term nature of agricultural burnings, we focused only on human short-term (24 h) exposure. Therefore, we set up AERMOD to calculate at every receptor the 24 h average pollutant concentration and to record only the maximum value obtained after one year of meteorological data (or the number of years of meteorological data availability). Finally, we selected all those receptors where pollutant concentration exceeded its respective threshold value specified in the NAAQS. The combination of all those receptors made up the influence area of agricultural burning.

The largest influence area is produced by the crop with the largest emission rate of the pollutant with the highest hazard to human health. According to Table 1 and Equation (2), for the case of sugarcane, the pollutant with the highest mass emission rate is PM_{10}. Although the PM_{10} emission factor for sugarcane is one of the lowest among the different crops listed in Table 1, sugarcane is the crop with the highest loading factor.

The hazard of a pollutant can be quantified as the inverse of its threshold value specified in the NAAQS. According to the American Conference of Governmental Industrial Hygienists (ACGIH), the threshold limit values (TLVs) are the maximum average airborne concentration of a hazardous material to which healthy adult workers can be exposed during an 8-h workday and 40-h workweek—over a working lifetime—without experiencing significant adverse health effects. They represent the opinion of the scientific community that exposure at or below the level of the TLV does not create an unreasonable risk of disease or injury [3]. Aiming to provide public health protection, including protecting the health of "sensitive" populations such as asthmatics, children, and the elderly, the environmental authority specifies those threshold limit values in the NAAQS for short time periods of exposure (3, 8, or 24 h depending of the pollutant) and for long time periods of exposure (one year) [14–24]. In this work, we only consider short-term exposition, as agricultural burnings are short-term events. Table 4 lists the USEPA maximum recommended values for short-term exposition.

Table 3. Sugarcane emission factors reported by several authors expressed as kg of pollutant emitted per Mg of sugarcane biomass burned. Highlighted boxes indicate the values used in this work.

Author, Year	BC *	TSP kg/Mg	PM_{10} kg/Mg	$PM_{2.5}$ kg/Mg	CO_2 g/kg	CO kg/Mg	NO kg/Mg
[22], 2011	-	-	2.3–2.5	-	-	30–41	-
[25], 2006	-	-	-	-	92	-	-
[26], 1996	-	4.31–4.64	4.51	4.19	-	55.83	3.18
[27], 2012	-	-	-	2.6 ± 1.6	1303 ± 218	65 ± 14	1.5 ± 0.4
[28], 2012	0.71 ± 0.22	-	-	2.49 ± 0.66	1255 ± 287	9.2 ± 3.3	-
[29], 2017	0.158	-	-	-	1791.94 ± 145.08	68.43 ± 16.23	1.63 ± 0.23
[30], 2018	-	3.27 ± 0.81	1.81 ± 0.14	1.19 ± 0.08	1618 ± 108	25.7 ± 2.04	-

* BC, Black carbon; -, no available data; Reference [22] corresponds to the USEPA recommended emission factors.

Table 4. Colombian NAAQS [31], emission rates for a burning rate of 1 ha/day, and risk indexes calculated for the case of the agricultural burning of sugarcane crops (j = sugarcane).

Pollutant	Colombian NAAQS Threshold Values ($\mu g/m^3$)	Colombian NAAQS Short Term Exposure (h)	Emission Rate (g/s)	$I_{i,j}$ (m^3/s)
TSP	300	24	2.47	8.2
PM_{10}	100	24	1.33	13.3
$PM_{2.5}$	50	24	2.23	44.6
CO	5000	8	21.83	0.4
NO_2	200	1	2.02	10.1

Aiming to identify the scenario that produces the largest agricultural burning influence area, we defined the risk index ($I_{i,j}$) for pollutant i generated from crop j, according to Equation (3). In this equation, AQ_i is the NAAQS threshold value for pollutant i. The crop and pollutant with maximum value for $I_{i,j}$ defines the largest area of influence. Table 4 shows the $I_{i,j*}$ obtained for the case of sugarcane crops ($j^* =$ sugarcane). It shows that $PM_{2.5}$ has the largest I_{i,j^*} and therefore it is the pollutant that defines the influence area for the burning of sugarcane crops.

$$I_{i,j} = \frac{E_{i,j}}{AQ_i} \tag{3}$$

3. Results

We studied the effect of the meteorology, emissions rate and size of the burning area. For illustrative purposes, we report results for the case of sugarcane burning. However, these results are valid for any crop.

3.1. The Effect of Meteorological Conditions on Pollutant Concentration

As a first step, we studied the effects of meteorology on the dispersion of pollutants. Arbitrarily, we kept the $PM_{2.5}$ emission rate constant at 1 g/s over a burning area of 1 ha. As expected, meteorology significantly affects $PM_{2.5}$ concentration at ground level. Figure 2 presents the daily maximum concentration obtained at any receptor over an extension of 10 km × 10 km, after considering the datasets of 1-h meteorological data listed in Table 2. This figure indicates that the meteorological data No. 2 (Minnesota) induced the highest level of pollutant concentration. This meteorology has an average temperature of 4.5 °C and wind speed of 2.7 m/s with no preferential wind direction. Besides low average wind speed, we could not identify a special characteristic of this meteorology that makes it the worst-case scenario. From now on, we only consider this meteorology as it constitutes the scenario that produces the highest concentrations.

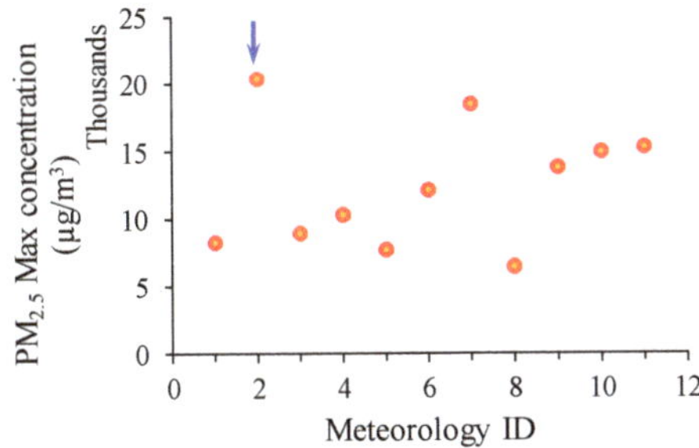

Figure 2. Maximum average daily $PM_{2.5}$ concentrations produced at any receptor over an extension of 10 km × 10 km by the burning of sugarcane biomass on a squared 1 ha area, after considering the datasets of 1-h meteorological data listed in Table 2. The arrow identifies meteorology No. 2 (Minnesota), which produced the highest $PM_{2.5}$ concentrations.

3.2. The Effect of Emission Rate on Pollutant Concentration

AERMOD has a linear response to changes in emissions. Aiming to confirm this expected behavior, a base emission of 1 g/s was used. This emission was multiplied by 0.1 and 10. We set these values as the new emissions rates and observed $PM_{2.5}$ concentrations nearby the emission source as predicted by AERMOD. Figure 3 compares $PM_{2.5}$ concentration obtained at every receptor in the base case scenario against the corresponding concentrations obtained with different emission rates. This comparison was performed in terms of normalized concentration, i.e., concentration divided by the emission rate. Figure 3 shows that all data points fall within the 45-degree line, regardless of the emission rate, confirming that, according to AERMOD, $PM_{2.5}$ concentration, at ground level, nearby the emission source is proportional to the emission rate.

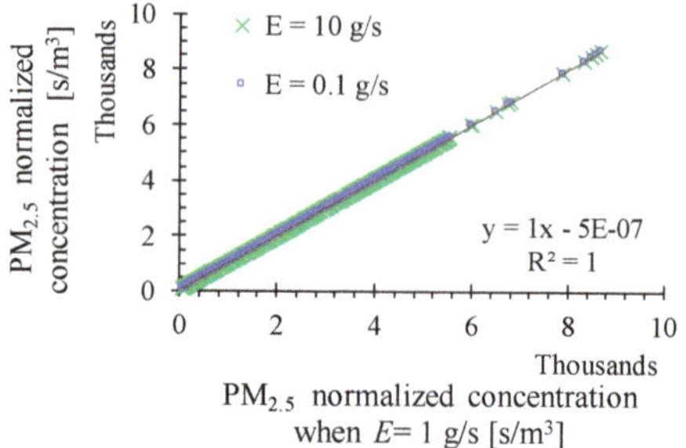

Figure 3. Comparison of the normalized PM$_{2.5}$ concentration at ground level obtained by AERMOD when the emission rate is 0.1 and 10 g/s against the normalized concentration when the emission rate is 1 g/s. Normalized concentration is obtained when the resulting concentration is divided by the emission rate in the source. Results were obtained for a burning area of 1 ha and the Minnesota meteorology dataset.

3.3. Determination of the Influence Area

As explained above, due to the short-term nature of the agricultural burning events, and because those events could happen at any time of the year, the determination of the influence area requires:

- The AERMOD determination of daily maximum concentrations, obtained at each receptor over the computational domain along the simulation time (1–5 years of 1 h meteorological data). The simulation should be carried out for the case of the riskiest pollutant at the emission rate calculated for that pollutant and crop of interest, in this case, PM$_{2.5}$ and sugarcane, respectively.
- A comparison of the obtained results against the threshold value specified in the NAAQS for short-term exposure to the riskiest pollutant, in this case, 50 μg/m^3 for 24 h of human exposure to PM$_{2.5}$.

Figure 4 shows the maximum daily PM$_{2.5}$ concentration obtained at each receptor located over a 10 km × 10 km region that surrounds a squared burning area of 1 ha with an hypothetical emission rate of E = 18.6 g/s. It shows that, due to the random wind direction, the influence area does not exhibit any regular shape. Therefore, for the case of agricultural burning, we redefined the influence area as the circle whose radio includes all areas where pollutant concentrations exceed the air quality standards defined by local environmental authorities.

Afterwards, we ran a set of cases changing the size of the area source from 1 m^2 to 20 ha and observed their resulting influence areas. As farmers partially control the burning rate by controlling the length and number of lines of starting fire fronts, we considered two alternatives:

- Farmers burn simultaneously the entire area, keeping the number of starting fire fronts per unit area constant. This alternative implies that, regarding of the burning area size, the burning event will be completed within the same period of time of the base case scenario (1 ha). It implies that the burning rate and the emission rate of pollutants per unit area remain constant. However, the total emission rate increases, with respect to the base case scenario, proportionally to the size of the area under consideration.
- Farmers burn sequentially one-unit area after another, increasing the duration of the burning event proportionally to the area size. This alternative implies that the total emission rate remains constant.

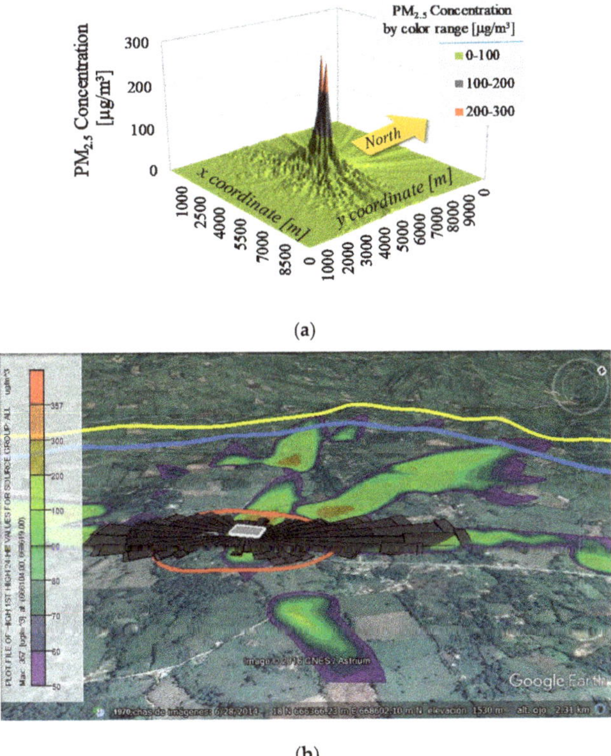

(a)

(b)

Figure 4. PM$_{2.5}$ ground concentration over a 10 km × 10 km region as result of the sugarcane burning on a square area of 1 ha at a burning rate of 18.6 g/s: (**a**) 3D representation of PM$_{2.5}$ concentration; and (**b**) 2D representation of PM$_{2.5}$ concentration. The lines in (**b**) represent circumferences centered in the emission source that limit the obtained influence area when the threshold value is 50 μg/m^3 (yellow), 100 μg/m^3 (blue) and 300 μg/m^3 (red).

In real practice, farmers burn with a combination of both alternatives and therefore we considered this third alternative in our simulations. In all cases, we reported the size of the resulting influence area as the radii of the resulting influence area minus the edge-size of the burning area.

We determined the size of the influence area generated by an area source of 1 ha, varying the emission rate. The obtained results are plotted in Figure 5a. It shows that the size of the influence area increases with the emission rate, following a logarithm profile. This profile crosses the area size axis at an emission rate of about 2 g/s. This means that farmers can burn at a rate smaller than this critical rate generating a negligible influence area. For the case of sugarcane, this value means a maximum burning rate of 1.5 ha/day.

When the emission rate remains constant, the size of the influence area remains constant regardless of the size of the burning area (Figure 5b). The burning of 1 m^2 at a given emission rate produces the same size of influence area as the burning of 1 ha at the same emission rate. The difference is that, under these circumstances, it takes 10^4 times longer to complete the burning task of 1 ha than of 1 m^2. This result implies that the size of the influence area is determined by the burnings near the edge of the area source. Aiming to observe the variation of these results with the orientation of the area source, we varied it from 0–170 degrees. The results are presented in Figure 5c. For the case of an area source of 1 ha, with an emission rate of 18.6 g/s, under all orientations that we ran, the influence area was ~2000 m considering $PM_{2.5}$ as the limiting pollutant. The influence area for PM_{10} and TSP (Total Suspended Particles) were ~1500 and 500 m, respectively.

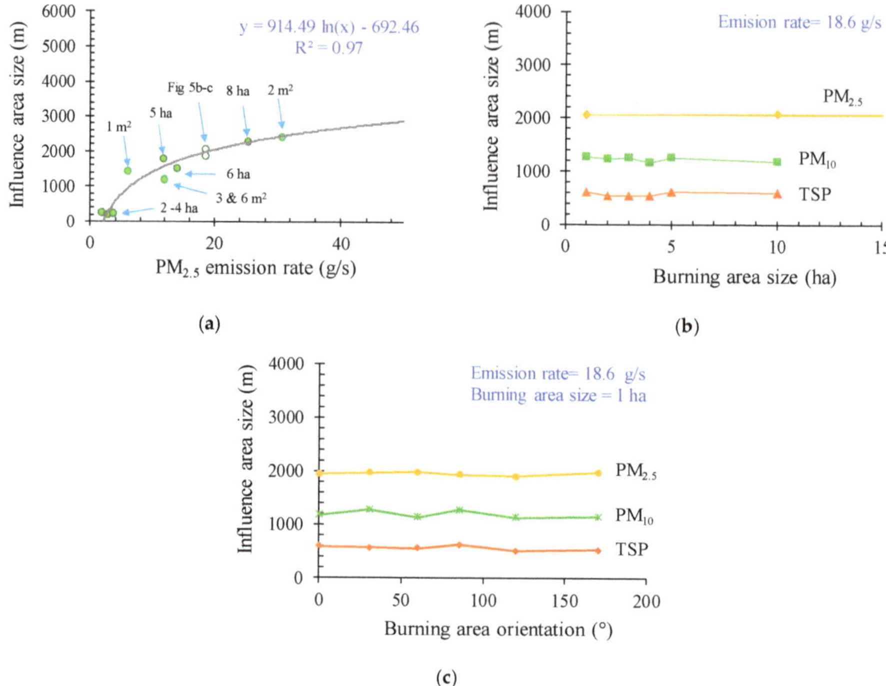

Figure 5. Size of the influence area generated by agricultural burning as function of: (**a**) emission rate considering different burning area sizes; (**b**) burning area keeping emission rate constant at 18.6 g/s; and (**c**) burning area orientation respect to north, keeping constant the emission rate at 18.6 g/s and the burning area at 1 ha.

Finally, we considered burning areas with shapes different from a square. The area of any irregular polygon can be constructed as the combination of multiple squares of different sizes. Using the principle of superposition, the influence area produced by the area source of irregular shape is the union of the influence areas generated by each independent squared area. Therefore, the influence area generated by the burning of crops cultivated on areas of any shape is the area band that surrounds the burning area. These results are independent of the type of crop or biomass being burned.

3.4. Recommendations for Policy Makers

In the light of this work, we suggest that policy makers interested in controlling the activity of agricultural burning and/or any open atmosphere biomass burning should be aware of:

- Open atmosphere biomass burning produces short- and long-term negative impacts on human health and the environment. Therefore, this practice should be controlled and eliminated as soon as possible. However, this activity is associated with important economic and social aspects that need to be considered. Therefore, environmental authorities, companies and the people that could be affected, should design in collaboration an action plan with a sustainable approach that ends with the elimination of this activity.
- Despite the efforts made by the scientific community to develop tools to assess accurately the impact of open biomass burning, several unresolved aspects and uncertainties remain related to:

(i) the amount of biomass burned per crop; (ii) the emission factors for the relevant pollutants per crop; (iii) the understanding and modeling of the pollutant dispersion phenomena; and (iv) secondary effects such as changes in atmospheric dynamics and alterations in the cloud formation processes.

- We used AERMOD to model the dispersion of the pollutants produced during agricultural burning events. This model is recommended by the USEPA for this type of applications. It means that, even though there could exist more accurate models for modeling agricultural burnings, AERMOD is the model that should be used for regulatory purposes, as it is well accepted by the scientific community and environmental authorities.
- Aiming to design public policies to control agricultural burning, the purpose of modeling the dispersion of the pollutants generated by this activity is to assess the environmental impact caused by the agricultural burning of any crop under a worst case but real scenario, considering all the possible pollutants that could be generated. In this regard, it is out of the scope of the present work to reproduce any measurements of pollutant concentration obtained nearby agricultural burning.

Based on the results obtained in this work, we propose that the environmental authorities:

- Limit any agricultural burning or any open atmosphere biomass burning to emissions rates smaller than 2.0 g/s calculated using Equation (2) and data in Table 1, for all pollutants regulated in the NAAQS. According to Figure 5a, this emission rate produces an influence area of negligible size. For the case of sugarcane, this counter-measure limits the burning rate to ~1,5 ha/day, which could be inappropriate for the current operation of the sugarcane industry.
- Determine the distance from the cultivated area to the location of the nearest household and use that distance as the size of an acceptable influence area. Then, use Figure 5a to determine the maximum allowable emission rate, which is directly related to the number of hectares that can be burned per day.
- The implementation of a burning management program that involves previous alternatives. This program divides the cultivated area in subareas, each of them with different distances to the nearest household. For each subarea, Figure 5a limits the maximum burning rate. Then, the burning management program establish the sequence that each area could be burned at the given burning rates. No two areas can be burned simultaneously.

4. Conclusions

Aiming to design public policies to control agricultural burning, we assessed the environmental impact generated by this activity. We used AERMOD to determine the concentration of the pollutants generated by this activity on the areas nearby the burning area (cultivated crop). We considered a wide range of meteorological conditions, burning rates, geometries, and sizes of burning areas.

The area influenced by a given agricultural burning is the largest area where the concentration of any of the pollutant under consideration exceeds its maximum threshold values established in the local atmospheric air quality standards (NAAQS). Results show that this area is a band around the cultivated area whose size increase with the emission rate of the riskiest pollutant (Figure 5a), but it does not depend on size of the burning area. The risk of a pollutant was quantified as the ratio of their emission rate to its threshold value established in the NAAQS.

The emission rate is proportional to the burning rate. As farmers control burning rate by controlling the length and number of starting flame fronts, we proposed the elaboration of a public policy to limit the burning rate in the way that no households are located within the resulting influence area. This proposal should be taken as a transitional stage towards a policy of no agricultural burnings in consideration of their adverse effects on the environment.

Author Contributions: D.F.P. Software, formal analysis, investigation, and original draft preparation. J.I.H. Conceptualization, formal analysis, writing—review and editing, and supervision.

Atmosphere **2019**, *10*, 312

Funding: This research was financed by the Mexican Council for Science and Technology (CONACYT) and by the Colombian Ministry of the Environment.

Acknowledgments: This study was partially financed by the Colombian Ministry of the Environment, the Mexican Council for Science and Technology (Consejo Nacional de Ciencia y Tecnología-CONACYT), and CAIA Engineering.

Conflicts of Interest: The authors declare no conflict of interest.

Appendix A Wind Roses Obtained for Each Set of Meteorological Data Used in This Study

Name	Year	Country	Wind Rose Diagram	Name	Year	Country	Wind Rose Diagram
San Diego	2009	USA	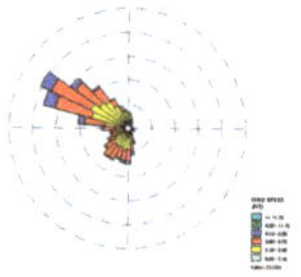	Zavala	2008–2012	USA	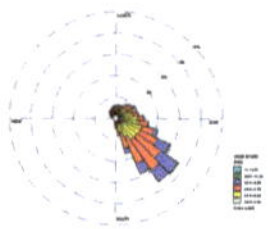
Minnesota	2008–2012	USA	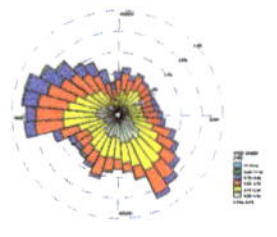	Pico	2008–2012	USA	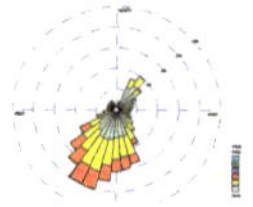
Texas	1990		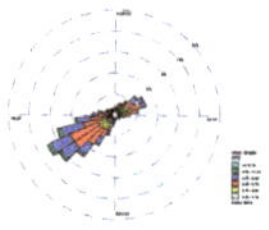	Descanso	2009	Colombia	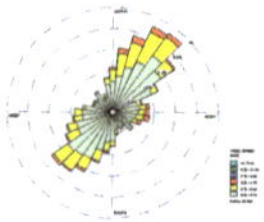
Michigan	2008–2012	USA	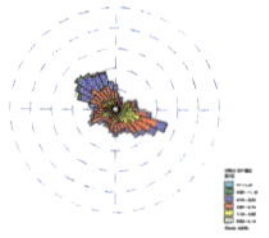	Cerro largo	2009	Colombia	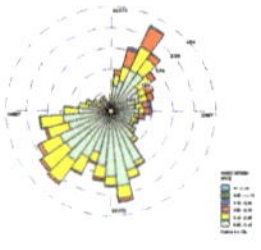
Alaska	1990	USA	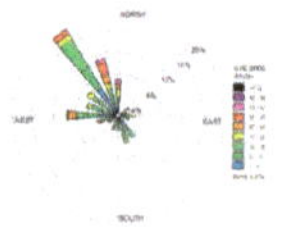	Rubiales	2013	Colombia	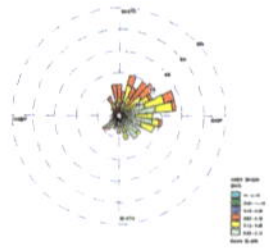
				Los Angeles	2012-2016	USA	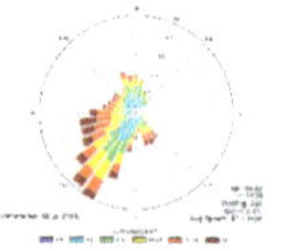

References

1. Holder, A.L.; Gullett, B.K.; Urbanski, S.P.; Elleman, R.; O'Neill, S.; Tabor, D.; Mitchell, W.; Baker, K.R. Emissions from prescribed burning of agricultural fields in the Pacific Northwest. *Atmos. Environ.* **2017**, *166*, 22–33. [CrossRef]

2. Ferreira, L.E.N.; Muniz, B.V.; Bittar, T.O.; Berto, L.A.; Figueroba, S.R.; Groppo, F.C.; Pereira, A.C. Effect of particles of ashes produced from sugarcane burning on the respiratory system of rats. *Environ. Res.* **2014**, *135*, 304–310. [CrossRef] [PubMed]

3. ACGIH. Threshold Limit Values (TLVs) and Biological Exposure Indices (BEIs). Appendix B. Signature publications. 2012. Available online: https://www.nsc.org/Portals/0/Documents/facultyportal/Documents/fih-6e-appendix-b.pdf (accessed on 4 April 2019).

4. Sahai, S.; Sharma, C.; Singh, S.K.; Gupta, P.K. Assessment of trace gases, carbon and nitrogen emissions from field burning of agricultural residues in India. *Nutr. Cycl. Agroecosyst.* **2011**, *89*, 143–157. [CrossRef]

5. Wieser, U.; Gaegauf, C.K. Nanoparticle emissions of wood combustion processes. In Proceedings of the First World Conference and Exhibition on Biomass for Energy and Industry, Sevilla, Spain, 5–9 June 2000; pp. 805–808.

6. Jimenez, J.; Wu, C.-F.; Claiborn, C.; Gould, T.; Simpson, C.D.; Larson, T.; Liu, L.-J.S. Agricultural burning smoke in eastern Washington—part I: Atmospheric characterization. *Atmos. Environ.* **2006**, *40*, 639–650. [CrossRef]

7. Badarinath, K.V.S.; Kumar Kharol, S.; Rani Sharma, A. Long-range transport of aerosols from agriculture crop residue burning in Indo-Gangetic Plains A study using LIDAR, ground measurements and satellite data. *J. Atmos. Solar-Terrestrial Phys.* **2009**, *71*, 112–120. [CrossRef]

8. Akagi, S.K.; Yokelson, R.J.; Wiedinmyer, C.; Alvarado, M.J.; Reid, J.S.; Karl, T.; Crounse, J.D.; Wennberg, P.O. Emission factors for open and domestic biomass burning for use in atmospheric models. *Atmos. Chem. Phys.* **2011**, *11*, 4039–4072. [CrossRef]

9. Chen, J.; Li, C.; Ristovski, Z.; Milic, A.; Gu, Y.; Islam, M.S.; Dumka, U.C. A review of biomass burning: Emissions and impacts on air quality, health and climate in China. *Sci. Total Environ.* **2017**, *579*, 1000–1034. [CrossRef] [PubMed]

10. Arbex, M.A.; Martins, L.C.; de Oliveira, R.C.; Pereira, L.A.A.; Arbex, F.F.; Cançado, J.E.D.; Saldiva, P.H.N.; Braga, A.L.F. Air pollution from biomass burning and asthma hospital admissions in a sugar cane plantation area in Brazil. *J. Epidemiol. Community Health* **2007**, *61*, 395–400. [CrossRef] [PubMed]

11. Mazzoli-Rocha, F.; Bichara Magalhães, C.; Malm, O.; Hilário Nascimento Saldiva, P.; Araujo Zin, W.; Faffe, D.S. Comparative respiratory toxicity of particles produced by traffic and sugar cane burning. *Environ. Res.* **2008**, *108*, 35–41. [CrossRef]

12. EPA. Solid Waste Disposal 2.5-1 Open Burning. 1995; 92. Available online: https://www3.epa.gov/ttn/chief/ap42/ch02/final/c02s05.pdf (accessed on 4 April 2019).

13. Prado, G.F.; Zanetta, D.M.T.; Arbex, M.A.; Braga, A.L.; Pereira, L.A.; de Marchi, M.R.; de Melo Loureiro, A.P.; Marcourakis, T.; Sugauara, L.E.; Gattás, G.J.; et al. Burnt sugarcane harvesting: Particulate matter exposure and the effects on lung function, oxidative stress, and urinary 1-hydroxypyrene. *Sci. Total Environ.* **2012**, *437*, 200–208. [CrossRef]

14. World Health Organization. *Evolution of WHO Air Quality Guidelines: Past, Present and Future*; WHO Regional Office: København, Denmark, 2017.

15. Wagner, J.; Naik-Patel, K.; Wall, S.; Harnly, M. Measurement of ambient particulate matter concentrations and particle types near agricultural burns using electron microscopy and passive samplers. *Atmos. Environ.* **2012**, *54*, 260–271. [CrossRef]

16. Mugica-Alvarez, V.; Hernández-Moreno, A.; Valle-Hernández, B.L.; Espejo-Montes, F.; Millán-Vázquez, F.; Torres-Rodríguez, M. Characterization and modeling of atmospheric particles from sugarcane burning in Morelos, Mexico. *Hum. Ecol. Risk Assess. Int. J.* **2017**, *7039*, 1–16. [CrossRef]

17. Hiscox, L.; Flecher, S.; Wang, J.J.; Viator, H.P. A comparative analysis of potential impact area of common sugar cane burning methods. *Atmos. Environ.* **2015**, *106*, 154–164. [CrossRef]

18. Carney, W.; Spicer, B.; Stegall, B.; Borel, C. Louisiana Smoke Management Guidelines for Sugarcane Harvesting. 2000. Available online: https://www.lsuagcenter.com/NR/rdonlyres/8AAEF1B2-EFA6-40A0-AC59-654C15894EE9/12567/smoke_management3.pdf (accessed on 4 April 2019).

19. Sornpoon, W.; Bonnet, S.; Kasemsap, P.; Prasertsak, P.; Garivait, S. Estimation of emissions from sugarcane field burning in Thailand using bottom-up country-specific activity data. *Atmosphere* **2014**, *5*, 669–685. [CrossRef]

20. Zhang, H.; Hu, J.; Qi, Y.; Li, C.; Chen, J.; Wang, X.; He, J.W.; Wang, S.X.; Hao, J.M.; Zhang, L.L.; et al. Emission characterization, environmental impact, and control measure of PM2.5 emitted from agricultural crop residue burning in China. *J. Clean. Prod.* **2017**, *149*, 629–635. [CrossRef]

21. Madriñán Palomino, C.E. Compilación y análisis sobre contaminación del aire producida por la quema y la requema de la caña de azúcar, *saccharum officinarum* L. en el valle geográfico del río cauca. 2002. Available online: http://bdigital.unal.edu.co/cgi/export/5039/ (accessed on 4 April 2019).

22. US EPA. AP 42, Fifth Edition, Volume I Chapter 13: Miscellaneous Sources. Section 13.2.1. 2011. Available online: https://www3.epa.gov/ttnchie1/ap42/ch13/ (accessed on 4 April 2019).

23. Fang, Z.; Deng, W.; Zhang, Y.; Ding, X.; Tang, M.; Liu, T.; Wang, X. Open burning of rice, corn and wheat straws: Primary emissions, photochemical aging, and secondary organic aerosol formation. *Atmos. Chem. Phys.* **2017**, *17*, 14821–14839. [CrossRef]

24. US EPA. NAAQS Table. Available online: https://www.epa.gov/criteria-air-pollutants/naaqs-table (accessed on 4 April 2019).

25. IPCC. 2006 IPCC Guidelines for National Greenhouse Gas Inventory. 2006. Available online: https://www.ipcc-nggip.iges.or.jp/public/2006gl/ (accessed on 4 April 2019).

26. Jenkins, B.M.; Turn, S.Q.; Williams, R.B.; Goronea, M.; Abd-el-Fattah, H.; Daniel Jones, A. Atmospheric Pollutant Emission Factors from Open Burning of Agricultural and Forest Biomass by Wind Tunnel Simulations. 1996 California Environmental Protection Agency. Available online: https://www.arb.ca.gov/ei/speciate/r01t20/rf9doc/a932-126_3.pdf (accessed on 4 April 2019).

27. França, D.A.; Longo, K.M.; Soares Neto, T.G.; Santos, J.C.; Freitas, S.R.; Rudorff, B.F.T.; Cortez, E.V.; Anselmo, E.; Carvalho, J.A., Jr. Pre-harvest sugarcane burning: Determination of emission factors through laboratory measurements. *Atmosphere* **2012**, *3*, 164–180. [CrossRef]

28. Hall, D.; Wu, C.-Y.; Hsu, Y.-M.; Stormer, J.; Engling, G.; Capeto, K.; Wang, J.; Brown, S.; Li, H.-W.; Yu, K.-M. PAHs, carbonyls, VOCs and PM2.5 emission factors for pre-harvest burning of Florida sugarcane. *Atmos. Environ.* **2012**, *55*, 164–172. [CrossRef]

29. Santiago-De la Rosa, N.; Mugica-Álvarez, V.; Cereceda-Balic, F.; Guerrero, F.; Yá-ez, K.; Lapuerta, M. Emission factors from different burning stages of agriculture wastes in Mexico. *Environ. Sci. Pollut. Res.* **2017**, *24*, 24297–24310. [CrossRef] [PubMed]

30. Mugica-Álvarez, V.; Hernández-Rosas, F.; Magaña-Reyes, M.; Herrera-Murillo, J.; Santiago-De La Rosa, N.; Gutiérrez-Arzaluz, M.; de Jesús Figueroa-Lara, J.; González-Cardoso, G. Sugarcane burning emissions: Characterization and emission factors. *Atmos. Environ.* **2018**, *193*, 262–272. [CrossRef]

31. Colombian Ministry of Environment. Resolution 2254, 1 November 2017. 1 November 2017. Available online: http://www.minambiente.gov.co/images/normativa/app/resoluciones/96-res%202254%20de%202017.pdf (accessed on 4 April 2019).

Article

Sensitivity of Nitrate Aerosol Production to Vehicular Emissions in an Urban Street

Minjoong J. Kim

Department of Environmental Engineering and Energy, Myongji University, Yongin 17058, Gyunggi, Korea; minjoongkim@mju.ac.kr

Received: 14 March 2019; Accepted: 18 April 2019; Published: 22 April 2019

Abstract: This study investigated the sensitivity of nitrate aerosols to vehicular emissions in urban streets using a coupled computational fluid dynamics (CFD)–chemistry model. Nitrate concentrations were highest at the street surface level following NH_3 emissions from vehicles, indicating that ammonium nitrate formation occurs under NH_3-limited conditions in street canyons. Sensitivity simulations revealed that the nitrate concentration has no clear relationship with the NO_x emission rate, showing nitrate changes of only 2% across among 16 time differences in NO_x emissions. NO_x emissions show a conflicting effect on nitrate production via decreasing O_3 and increasing NO_2 concentrations under a volatile organic compound (VOC)-limited regime for O_3 production. The sensitivity simulations also show that nitrate aerosol is proportional to vehicular VOC and NH_3 emissions in the street canyon. Changes of VOC emissions affect the nitrate aerosol and HNO_3 concentrations through changes in the O_3 concentration under a VOC-limited regime for O_3 production. Nitrate aerosol concentration is influenced by vehicular NH_3 emissions, which produce ammonium nitrate effectively under an NH_3-limited regime for nitrate production. This research suggests that, when vehicular emissions are dominant in winter, the control of vehicular VOC and NH_3 emissions might be a more effective way to degrade $PM_{2.5}$ problems than the control of NO_x.

Keywords: Urban pollution; Street canyon; Nitrate aerosol; CFD; Air quality

---

1. Introduction

Nitrate aerosol is a fine particulate matter ($PM_{2.5}$) component produced from the reaction of gas-phase nitrate (nitric acid; HNO_3) and ammonia (NH_3). During a haze event, nitrate aerosol often contributes to the total observed particle mass by as much or more than the organic fraction across East Asia [1–4]. As severe haze events have increased across East Asia [5], nitrate contributions to $PM_{2.5}$ mass have also increased in polluted urban areas [1,6].

The production of nitrate aerosol in urban areas is affected by vehicular emissions such as nitrogen oxides (NO_x = NO + NO_2) and NH_3 [7–9]. These vehicular emissions are highly concentrated and are transported by turbulence from the complex geometry of buildings, causing steep gradients of pollutant concentrations [10–13]. Nitrate aerosol chemistry is highly nonlinear and follows the concentrations of precursor gases and humidity, so the formation and distribution of nitrate aerosol are not spatially uniform [14,15]. Most modeling studies on nitrate aerosols are based on regional or global air-quality models that have clear limitations due to their coarse resolution [16–18]. Therefore, to accurately investigate nitrate formation in urban areas, fine-scale simulations that can conserve highly concentrated emission plumes and turbulence are necessary.

Meanwhile, policies regarding vehicular nitrate control focus on reduction of NO_x (e.g., through banning diesel vehicles) [19,20], based on studies of regional or global air-quality modeling. However, studies from observation campaigns have often reported that other vehicle emissions are much more important than NO_x emissions for nitrate production in urban areas [21–23]. Link et al.

investigated the secondary formation of ammonium nitrate from vehicle exhaust using sampling and laboratory experiments in the Seoul Metropolitan Region (SMR) [21]. They found that the secondary production of ammonium nitrate from diesel exhausts is much lower than that from gasoline. They concluded that the NH_3 source from gasoline vehicles could be more important than NO_x emissions, indicating that SMR is an NH_3-limited environment. Wen et al. studied nitrate formation during a severe $PM_{2.5}$ pollution period and reported that high NH_3 concentrations in the early mornings significantly accelerated the formation of fine particulate nitrate [23]. In addition, they found that the increased rate of nitrate aerosol had a strong positive correlation with ozone (O_3) concentrations at night, indicating the essential role of oxidants in nitrate formation. Studies also reported that volatile organic compound (VOC) concentrations control nitrate formation by affecting the O_3 levels [22,24]. These studies indicate that nitrate aerosols are affected by a complex chemical condition that involves NO_x, NH_3, and oxidants, whereas policy tends to focus exclusively on NO_x control. Thus, understanding the favorable conditions for nitrate formation in urban areas is crucial for the design of air-quality policies.

Thus, this study investigates the distribution of nitrate aerosols and the favorable conditions for nitration formation in urban streets using a microscale coupled computational fluid dynamics (CFD)–chemistry model that can reproduce the turbulence from complex building geometries. Sensitivity simulations were conducted to examine the sensitivity of emissions to nitrate production by changing the emissions of precursor gases for nitrate aerosols and oxidants. The sensitivity simulation results reveal what significant factors lead to nitrate aerosol problems in urban streets.

2. Model Description and Simulation Set-Up

2.1. Model Description

A coupled CFD–chemistry model was used based on that proposed by Kim et al. [25]. The CFD model is based on the Reynolds-averaged Navier–Stokes equation (RANS) model and assumes a three-dimensional (3-D), nonrotating, nonhydrostatic, and incompressible airflow system [26]. This model was previously used to examine the flow and dispersion of both reactive gas pollutants [25,27] and reactive aerosol pollutants [25,28].

The model's chemical mechanism includes a full tropospheric NO_x–O_x–VOC chemistry scheme from a global 3-D chemical transport model (GEOS-Chem V11-1) [29]. GEOS-Chem was initially developed to solve global air chemistry issues; however, application of the GEOS-Chem model has now been extended to the regional scale. The GEOS-Chem model can successfully explain urban air quality, including cases of severe haze over East Asia [1,30,31]. The chemical scheme includes 140 species and 393 reactions, among which 61 reactions are photochemical. Among the 140 species simulated in the chemistry module, the CFD model transports 65 chemical tracers. Radical species with very short chemical lifetimes are not transported. Photolysis rate coefficients are calculated using the Fast-JX radiative transfer model [32,33].

The model also calculates aerosols that include sulfate, nitrate, ammonium, black carbon, and organic carbon [34,35]. Sulfate formation generally occurs via two pathways: the gas-phase oxidation of SO_2 by OH and the aqueous-phase oxidation of SO_2 by ozone and hydrogen peroxide. The CFD model only accounts for the gas-phase oxidation of SO_2 by OH because it lacks an atmospheric physics module that simulates hydrometeors such as clouds and rain.

Nitrate and ammonium aerosol were calculated by partitioning the total NH_3 and HNO_3 between the gas and aerosol phases. We used the ISORROPIA-II model as a thermodynamic equilibrium model for aerosol partitioning [36,37] and employed it to calculate the thermodynamic equilibrium of a K^+–Ca^{2+}–Mg^{2+}–NH_4^+–Na^+–SO_4^{2-}–NO_3^-–Cl^-–H_2O aerosol system based on the NH_3, HNO_3, and SO_4^{2-} concentrations.

The model also includes the production of HNO_3 via heterogeneous chemistry between aerosols and gases following Jacob (2000) [38]. The reactions that contribute to HNO_3 production by heterogeneous chemistry can be written as:

$$2\,NO_2 \rightarrow HNO_3 + HONO \tag{R1}$$

$$NO_3 \rightarrow HNO_3 \tag{R2}$$

$$N_2O_5 \rightarrow 2\,HNO_3. \tag{R3}$$

The uptake coefficients, γ, of R2 and R3 were 10^{-4} and 0.1, respectively, following Jacob (2000) [38]. For R4, γ was set to 0.01 as suggested by Zhang et al. (2012) and Walker et al. (2012) [39,40]. N_2O_5 is essential for night-time nitrate aerosol chemistry; the production and loss reactions of N_2O_5 can be written as:

$$NO_2 + NO_3 + M \rightarrow N_2O_5 + M \tag{R4}$$

$$N_2O_5 + M \rightarrow NO_2 + NO_3 + M \tag{R5}$$

$$N_2O_5 + h\nu \rightarrow NO_3 + NO_2. \tag{R6}$$

The simulation of carbonaceous aerosols follows the GEOS-Chem model [35]. The primary carbonaceous aerosol follows the passive tracer without any chemical reactions. However, the model resolves primary Black Carbon (BC) and Organic carbon (OC) with a hydrophobic and a hydrophilic fraction for each (i.e., making four aerosol types) for deposition processes. All sources emit hydrophobic aerosols that then become hydrophilic with an e-folding time of 1.2 days, as per Cooke et al. (1999) [41]. Although secondary organic aerosol (SOA) chemistry is not considered in the model, we treat boundary inflow and the transport of SOAs. However, the model does not account for either dust or sea-salt aerosol.

The dry deposition of gases and aerosols was simulated with a standard big-leaf resistance-in-series model [42]. The model accounts for the dry deposition of 46 species, including aerosols. The CFD model has no atmospheric physics module that simulates hydrometeors (e.g., clouds and rain); thus, wet deposition was not calculated, following Kim et al. (2012) [25].

2.2. Simulation Set-Up

We assumed a street canyon selected for a simulation located in SMR, one of the most polluted cities, to investigate nitrate formation in urban streets. The domain size was $120 \times 80 \times 100$ m in the x, y, and z directions, respectively. The grid intervals in all directions were 2 m, and the building height was 20 m with unified aspect ratios for all street canyons. Figure 1 shows the detailed structure of the simulation domain.

First, we set a control run (CNTL hereafter) with standard emissions. We estimated the pollutant emissions from average traffic volume in 2017 obtained from the Traffic Monitoring System (http://www.road.re.kr), which provides traffic statistics in SMR. On that basis, the daily traffic volume for each street was assumed to be 15,130 vehicles day^{-1}. The monthly averaged diurnal variation in traffic volume was also obtained from the Traffic Monitoring System. Vehicular emissions were computed using the emission factor from the Clean Air Policy Support System (CAPPS) emission inventory [43] and calculated by multiplying the mean ratios of vehicle sizes in Korea. The emission factors following vehicle size and the ratios of vehicle size are summarized in Tables 1 and 2, respectively. The calculated averaged emissions per vehicle were 0.10 g km^{-1}, 0.15 g km^{-1}, 0.015 g km^{-1}, and 0.011 g km^{-1} for NO_x, CO, VOC, and NH_3, respectively. NO_x emissions were separated into NO and NO_2 emissions at a 10:1 ratio by volume [44]. Total VOC emissions were further speciated using the method developed by EMEP/EEA (2016) [45]. Table 3 lists the emission rate and ratios of speciated VOC in the CNTL simulation. All vehicular emissions were emitted at z = 1 m from 4 m wide area sources located at the center of the streets. Other emissions in the model domain were not considered in the simulation. Previous studies have reported that the largest proportion of emissions in SMR is vehicular emissions [43,46,47].

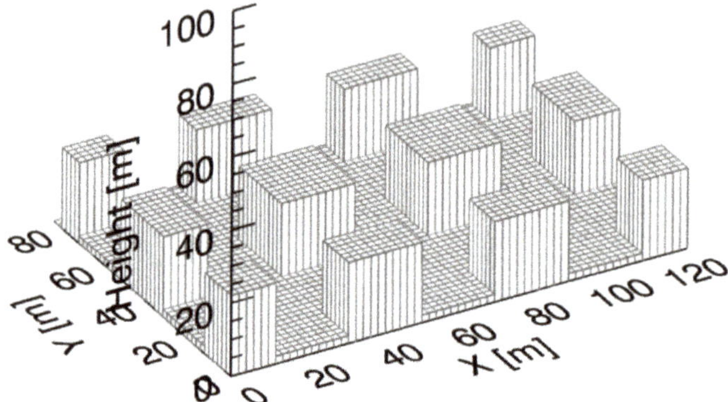

Figure 1. Schematic diagram of the coupled computational fluid dynamics (CFD)–chemistry simulation domain for the control run (CNTL) simulation.

Table 1. Emission estimates from vehicles used in the coupled CFD–chemistry model simulations for the CNTL simulation. The text discusses details about the species emission estimates.

[g km^{-1}]	CO	NO$_x$	VOC	NH$_3$
PC (diesel)	0.010	0.041	0.006	0.002
PC (gasoline)	0.085	0.004	0.002	0.023
PC (gas)	0.104	0.002	0.001	0.001
LDV (diesel)	0.087	0.063	0.009	0.002
LDV (gas)	0.163	0.026	0.002	0.002
HDV (diesel)	0.543	0.959	0.104	0.002
HDV (gas)	0.457	0.045	0.009	0.003
Bus (diesel)	1.850	0.613	0.150	0.002
Bus (gas)	1.736	0.346	0.191	0.000

Table 2. Ratios of vehicles following size and fuel type used in the CNTL simulation obtained from the Traffic Monitoring System.

[%]	Passenger Cars	Light-Duty Vehicles	Heavy-Duty Vehicles	Bus
Diesel	20.6	17.6	7.3	2.2
Gasoline	43.3	-	-	-
Gas	8.3	0.9	0.3	0.4
Total	72.2	18.5	7.6	2.6

For meteorological conditions, we used values for the SMR in winter that provided favorable conditions for nitrate formation [48]. To simulate diurnal changes in buoyancy and their effects on transport and chemical reaction rates in the model, we used the hourly temperature and relative humidity obtained from the Seoul station of the Korea Meteorological Administration (http://web.kma.go.kr/eng/). Table 4 shows the hourly temperature and relative humidity used in this study. The wind speed on the rooftop was assumed to be the observed seasonal mean value for SMR, which is 2.4 m s^{-1}. The wind direction is westerly (most frequently observed in winter) and is perpendicular to the street canyon [49]. The ambient wind speed and direction were fixed during a one-day simulation. The following vertical profiles of the wind, turbulent kinetic energy (TKE), and TKE dissipation rate were imposed:

$$U(z) = \frac{u_* \cos\theta}{\kappa} ln\left(\frac{z}{z_0}\right) \tag{1}$$

$$V(z) = \frac{u_* \sin\theta}{\kappa} ln\left(\frac{z}{z_0}\right) \tag{2}$$

$$W(z) = 0 \tag{3}$$

$$k(z) = \frac{u_*^2}{C_\mu^{1/2}}\left(1 - \frac{z}{\delta}\right)^2 \tag{4}$$

$$\varepsilon(z) = \frac{C_\mu^{3/4} k^{3/2}}{\kappa z} \tag{5}$$

Here, u_*, z_0, and κ represent the friction velocity, roughness length (=0.05 m), and von Karman constant (=0.4), respectively; C_μ is an empirical constant (=0.0845), and θ is the wind direction. The surface and top boundary pressures in the model were assumed to be 1013.15 hPa and 993.72 hPa, respectively.

Table 3. Emission rates per vehicle and ratios of speciated volatile organic compound (VOC) following the method developed by EMEP/EEA (2016).

Tracer Name	Formula	Emission Rate [mg km^{-1}]	Ratio [%]
ALK4	$\geq C_4$ alkanes	3.60	24.0
ISOP	$CH_2 = C(CH_3)CH = CH_2$	-	-
ACET	$CH_3C(O)CH_3$	0.44	2.9
MEK	$RC(O)R$	0.18	1.2
ALD2	CH_3CHO	0.98	6.5
RCHO	CH_3CH_2CHO	1.89	12.6
MVK	$CH_2 = CHC(=O)CH_3$	-	-
MACR	$CH_2 = C(CH_3)CHO$	-	-
PRPE	$\geq C_3$ alkenes	2.58	17.2
C3H8	C_3H_8	0.02	0.1
CH2O	$HCHO$	1.80	12.0
C2H6	C_2H_6	0.05	0.3
Unspeciated	-	-	23.2

Table 4. Diurnal variations of the observed hourly surface temperature and relative humidity used in this model.

Hour	01	02	03	04	05	06	07	08	09	10	11	12
Temperature [K]	−3.1	−3.4	−3.6	−4.0	−4.4	−4.6	−4.8	−5.1	−4.5	−2.7	−0.8	0.6
Relative Humidity [%]	37.7	38.3	37.8	36.8	37.4	36.9	37.5	37.6	32.2	26.4	21.3	17.9
Hour	13	14	15	16	17	18	19	20	21	22	23	24
Temperature [K]	1.4	2.2	2.7	2.6	1.6	0.3	−0.5	−1.2	−1.7	−2.1	−2.4	−2.8
Relative Humidity [%]	16.5	14.4	13.7	14.0	18.0	21.6	26.0	30.2	32.6	34.5	34.7	35.8

For the rooftop boundary conditions of the species, we used a reanalysis dataset from the Monitoring Atmospheric Composition and Climate (MACC) project in SMR with 6 h diurnal variations [50]. In addition, we used the boundary conditions of the species from the GEOS-Chem simulation with 6 h diurnal variations at SMR in winter for the species that were not provided in the MACC reanalysis dataset [51]. The initial conditions of the species were assumed as the concentrations of the boundary conditions in the first step. We conducted 48 h model simulations for each case: the first 24 h for the model spin-up, and the results from the last 24 h were used. The chemical and dynamical time steps were 1 min and 1 s, respectively.

Sensitivity simulations were conducted to examine the effect of emissions on nitrate aerosols in urban streets. Twelve sensitivity simulations were set by changing the vehicular emissions of NO_x, VOC, and NH_3. Each simulation was conducted with different emissions by multiplying the original emission by 0.25, 0.5, 2.0, and 4.0 for each species. We named the sensitivity simulations "species name" × "multiplying factor." For example, simulations named $NO_x \times 0.25$, $NO_x \times 0.5$, $NO_x \times 2$, and $NO_x \times 4$ indicate multiplying the vehicular NO_x by 0.25, 0.5, 2.0, and 4.0, respectively, while other emissions were fixed.

2.3. Model Validation

The coupled CFD–chemistry model in this study was thoroughly validated for the flow and dispersion of passive tracers in street canyons by comparing the results from this model with those from a wind tunnel, an idealized numerical study, and fluid experiments [25,27,52]. Park et al. (2015) found good agreements when comparing the model with wind tunnel data and experimental data by implementing improved wall functions for the momentum and thermodynamic energy equations in the CFD model to more accurately represent the effects of solid–wall boundaries [27].

The coupled CFD model also evaluated the dispersion of reactive pollutants compared with idealized simulations and field campaigns [25]. Kim et al. (2012) applied the coupled CFD–chemistry model to simulations using the same building configuration as in Baker et al. (2004) [25,53]. Their results showed that the concentrations of NO_x and O_3 have a pattern and magnitude consistent with the simulated concentrations by Baker et al. (2004) under steady-state O_3–NO–NO_2 photochemistry. Kim et al. (2012) reproduced reactive pollutants on Dongfeng Middle Street, Guangzhou, China, using a full tropospheric NO_x–O_x–VOC chemistry scheme and compared the results to a field campaign by Xie et al. (2003) [13]. The coupled model, with the full photochemical mechanism, also successfully captured the time variation in the observed CO concentrations for both upwind and downwind sites in the Dongfeng Street canyon. However, the coupled model overestimated NO_x concentrations compared to observations by Xie et al. (2003) due to estimating excessive NO emissions from traffic volume, implying the necessity of utilizing an accurate emissions inventory [13].

The coupled CFD model has also been used to evaluate the dispersion of reactive aerosol in street canyons [28]. Kim et al. (2019) evaluated the composition of PM_1 in summer and winter in a street canyon by comparison with the field campaign in Elche, Spain, by Yubero et al. (2015) [28,54]. The model generally captured seasonal variations of PM_1 in the street canyon. We evaluated the seasonal variation in nitrate concentration by comparing our model results to those of Yubero et al. (2015) [54]. Four simulations were conducted to represent the four seasons (spring, summer, autumn, and winter) in Elche, Spain. The street was approximately 7 m wide and surrounded by buildings that were approximately 25 m in height. The domain size was 20 m × 40 m × 50 m, and the number of grid points was 42 × 82 × 52. The meteorological conditions used were the observed seasonal mean values during the campaign periods. Pollutant emissions were estimated from traffic volume obtained from the Elche Traffic Office [54]. Vehicular emissions were computed using Spain's emission rates in EMEP/EEA (2016). The detailed model configuration generally followed that of Kim et al. (2019) [28].

Figure 2 shows the observed and simulated nitrate concentrations by season. The observed nitrate concentrations were highest in winter and lowest in summer due to the thermal evaporation of nitrate aerosols, showing 1.3 and 0.3 µg m^{-3}, respectively. The observed nitrate concentrations in spring and autumn were 0.4 and 0.5 µg m^{-3}, respectively, falling between those in winter and summer and indicating the dominant effect of temperature on nitrate aerosol. The simulated nitrate concentrations were 1.0, 0.8, 1.0, and 1.4 µg m^{-3} in spring, summer, autumn, and winter, respectively, showing clear seasonal variation. The simulated concentrations reproduce the observed seasonal variation in nitrate concentration. However, the simulated nitrate concentration in summer was twice that of the observed magnitude, indicating the weaker thermal evaporation of ammonium nitrate predicted by the model. The model captured the magnitude of the observed nitrate in winter located within the standard deviation of the observed nitrate concentration. These results indicate that the model successfully calculated the nitrate aerosol in cold environments.

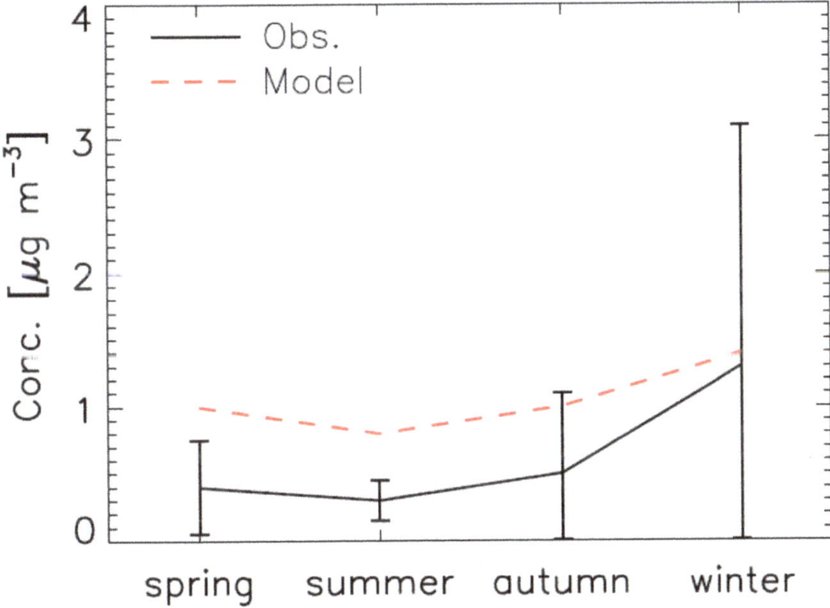

Figure 2. Comparisons of nitrate concentrations in the spring, summer, autumn, and winter cases between this work and the previous results of Yubero et al. (2015) (see Figure 1 in Yubero et al.). The values are averaged for the analysis period at the sample station. The black solid line is the observed concentrations and the red dashed line is the simulated concentrations. The error bars indicate the standard deviations of the observed nitrate aerosol in each season.

3. Results

Before investigating the production and sensitivity of nitrate aerosol, we checked the precursor gases of nitrate aerosol and oxidant concentrations that affect nitrate aerosol chemistry. Figure 3a indicates the meridionally averaged NO$_x$ concentration in the domain. The NO$_x$ concentration is highest at the street surface level, indicating that vehicular NO$_x$ emission is trapped by the canyon vortex in the street canyon. The NO$_x$ concentration reached 100 ppbv, ten times higher that outside the canyon, showing a steep gradient of NO$_x$ concentration in the street canyon. Note that concentrations in the three street canyons have slightly different values and dispersion patterns owing to different vortex patterns under non-infinite consecutive 3-D street canyons following the dispersion rates of TKE [26]. Figure 3b shows the spatial distribution of O$_3$ concentration. The O$_3$ concentration was lowest at the surface, showing a negative correlation with the NO$_x$ concentration. The O$_3$ concentrations outside and inside the street canyon were 38 and 11 ppbv, respectively, suggesting NO$_x$ titration in the street canyon. This distribution of O$_3$ and NO$_x$ fits the general dispersion pattern in the street canyon reported in previous studies [25,27]. Figure 3c displays the HNO$_3$ concentration, which reached 3.8 ppbv in the street canyon. The HNO$_3$ concentration in the street canyon was higher than that outside the street canyon, indicating the oxidation of HNO$_3$ from vehicular NO$_x$. However, the HNO$_3$ concentration at the surface was the lowest, even though the NO$_x$ concentration was highest at the surface. The low level of HNO$_3$ at the surface was caused by low O$_3$ concentrations at the surface under VOC-limited conditions, suppressing the production of OH and HNO$_3$. Figure 3d displays the NH$_3$ concentrations; the daily averaged NH$_3$ concentration reached 2.3 ppbv, with the highest value at the surface. The dispersion pattern of NH$_3$ was similar to that of NO$_x$, indicating the high impact of vehicular emissions.

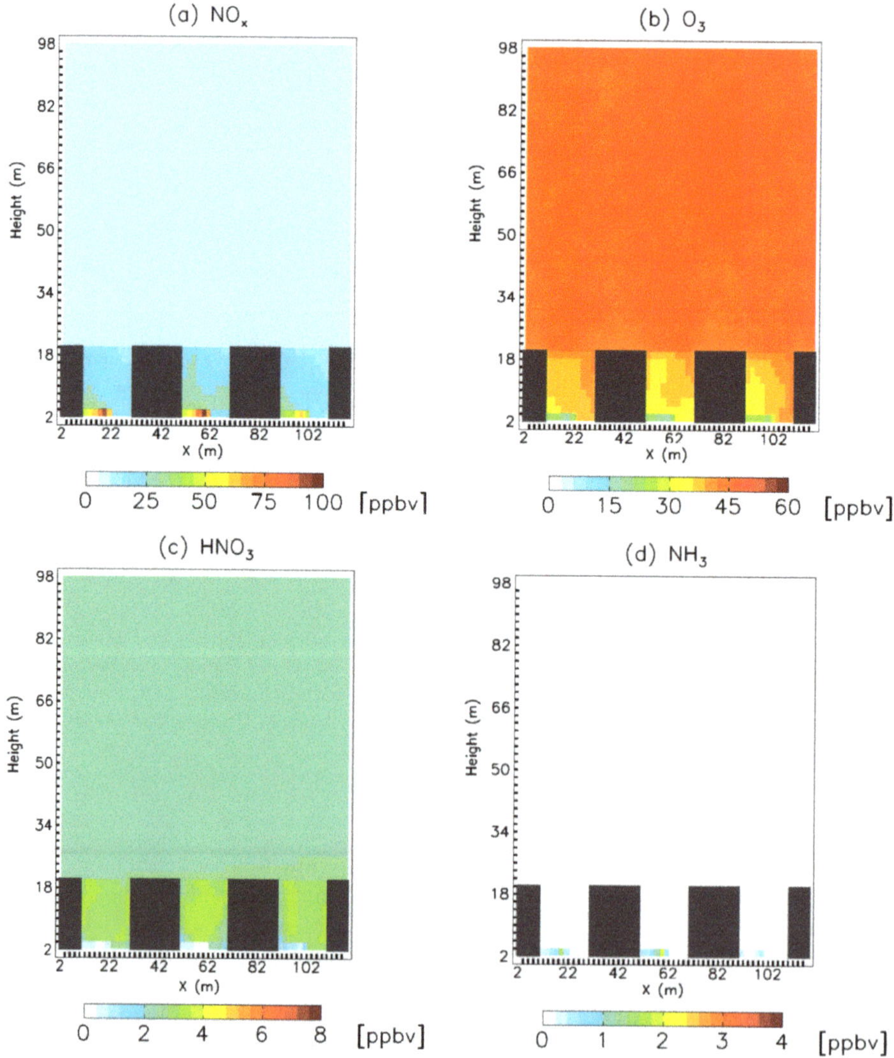

Figure 3. Distributions of the daily average concentrations of (**a**) NO$_x$, (**b**) HNO$_3$, (**c**) O$_3$, and (**d**) NH$_3$ (ppbv) in the CNTL simulation.

Figure 4a shows the nitrate concentrations in the street canyon, which are higher than those outside the canyon, showing values of up to 11.4 $\mu g\ m^{-3}$. The high concentration of nitrate aerosol is due to the high HNO_3 concentration from vehicular NO_x in the street canyon. However, the spatial pattern of nitrate aerosol in the street canyon differs from that of HNO_3, because NH_3 is a precursor gas of ammonium nitrate. The nitrate aerosol was highest at the surface following vehicular NH_3. These high correlations between NH_3 and nitrate aerosol indicate that ammonium nitrate forms under NH_3-limited conditions in the street canyon. Figure 4b shows the ammonium concentration in the street canyon; the spatial distribution of ammonium aerosol is similar to that of nitrate, implying that most ammonium aerosols combine with nitrate aerosols in winter. The maximum concentration of ammonium aerosol was 4.7 $\mu g\ m^{-3}$, and the spatial patterns of ammonium also indicate a low concentration of ammonium sulfate in winter. Note that the vehicular emission rate of SO_2 was very low, implying that ammonium sulfate inside the street canyon might also be low [55]. The sum of the ammonium and nitrate concentrations was 16.1 $\mu g\ m^{-3}$, higher than the air quality guidelines set by the World Health Organization (WHO, 10 $\mu g\ m^{-3}$) and 46% of the WHO Interim Target-1 (35 $\mu g\ m^{-3}$), indicating the hazardous effect of ammonium nitrate aerosols on pedestrians [56]. Considering that nitrate chemistry is highly nonlinear, this cannot be resolved and may lead to uncertainty in regional models due to their coarser resolution.

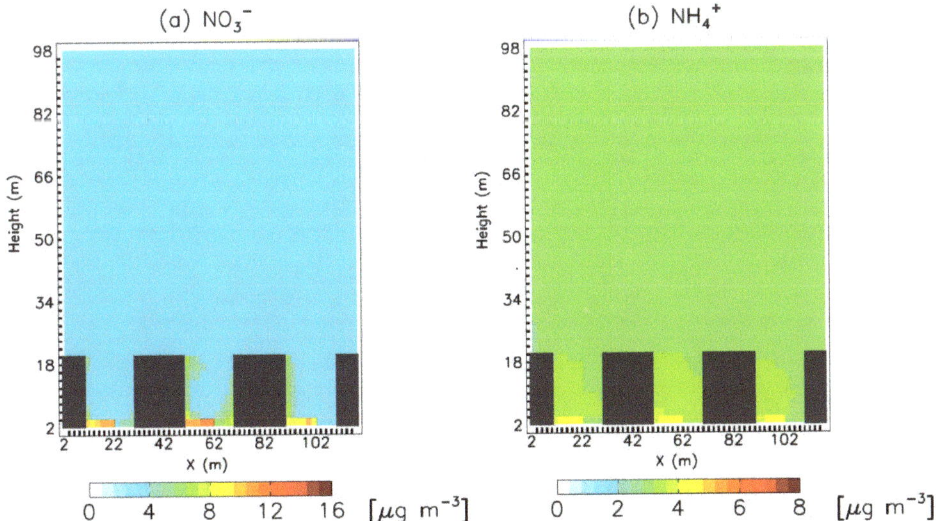

Figure 4. Distribution of the daily average concentrations of (**a**) nitrate and (**b**) ammonium ($\mu g\ m^{-3}$) in the CNTL simulation.

We investigated the sensitivity of nitrate aerosol production to NO_x emission. Figure 5a shows the average nitrate concentration in the street canyon (i.e., below 20 m) by following vehicular NO_x emission changes to investigate the sensitivity of nitrate aerosol production to the vehicular NO_x emission rate. Surprisingly, the nitrate concentration did not show a clear relationship with the NO_x emission rate. Nevertheless, the change in nitrate concentrations was at most 2% compared to the standard simulation, and the average nitrate concentration was highest in the CNTL and lowest in the $NO_x \times 0.25$ simulations, indicating the nonlinearity of the nitrate aerosol production to the NO_x emission rate. These results contradict the conventional belief that high NO_x emissions from vehicles can cause nitrate aerosol air quality problems.

Figure 5b,d display the average HNO_3, O_3, and NO_2 concentrations, respectively, in the street canyon according to the sensitivity simulations. The HNO_3 concentration in the street canyon shows changes similar to those of nitrate aerosol, indicating that the changes in nitrate aerosol in the sensitivity simulations are closely related to the changes in HNO_3 (Figure 5b). The O_3 concentration decreases as the NO_x emissions increase because of NO_x titration (Figure 5d). Note that O_3 formation falls under a VOC-limited regime in the street canyon due to vehicular emissions. A low concentration of O_3 prevents the conversion of NO_2 to HNO_3 through either photochemical production during the daytime due to inhibited OH production and heterogeneous nitrate production at night. The NO_2 (the precursor gas of HNO_3), concentration in the street canyon is proportional to the NO_x emissions because it affects the direct NO_2 emissions from vehicles and the reduction in photodissociation of NO_2 under low O_3 concentrations (Figure 5c). High NO_2 creates suitable conditions for HNO_3 formation, compensating for the effect of decreased O_3 on HNO_3. Thus, HNO_3 and nitrate aerosols have no clear relationship with NOx emissions and only undergo small changes. These results imply that NO_x emission controls cannot improve $PM_{2.5}$ levels in urban street conditions.

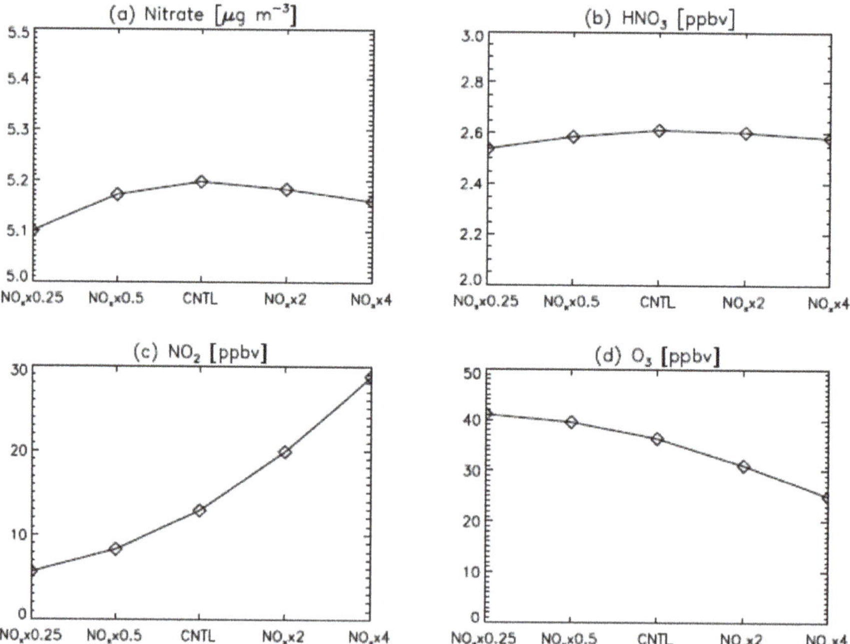

Figure 5. Average (**a**) nitrate, (**b**) HNO_3, (**c**) NO_2, and (**d**) O_3 concentrations in the street canyon (i.e., below 20 m) following a change in the emission rate of vehicular NO_x.

In addition, we estimated the sensitivity of nitrate aerosol production to VOC emissions. Figure 6a shows the average concentration of nitrate aerosol in the street canyon following vehicular VOC emission changes; nitrate concentrations are proportional to VOC emissions. The average nitrate concentration in the VOC × 0.25 simulation was 8% lower than that of the CNTL simulation, indicating the higher sensitivity of VOC emissions to nitrate aerosols compared with that of NO_x emissions. The nitrate aerosols in VOC × 0 only showed a 12% difference with those of the CNTL simulation, which implies the large impact of the boundary condition on nitrate formation. The changes in HNO_3 concentration follow the changes in nitrate concentration, implying that the former is caused by the latter (Figure 6b). Reducing VOC emissions drives a decrease in both NO_2 and O_3 concentrations, creating unfavorable conditions for HNO_3 production, in contrast to the effect of NO_x emissions (Figure 6c,d). These results are consistent with a previous box modeling study, which suggests that increases in VOC emissions induce nitrate production via O_3 increases under a VOC-limited regime for O_3 production [57]. Considering that megacities are generally under a VOC-limited regime [58,59], VOC emission control can improve both O_3 control and $PM_{2.5}$ control in urban street canyons.

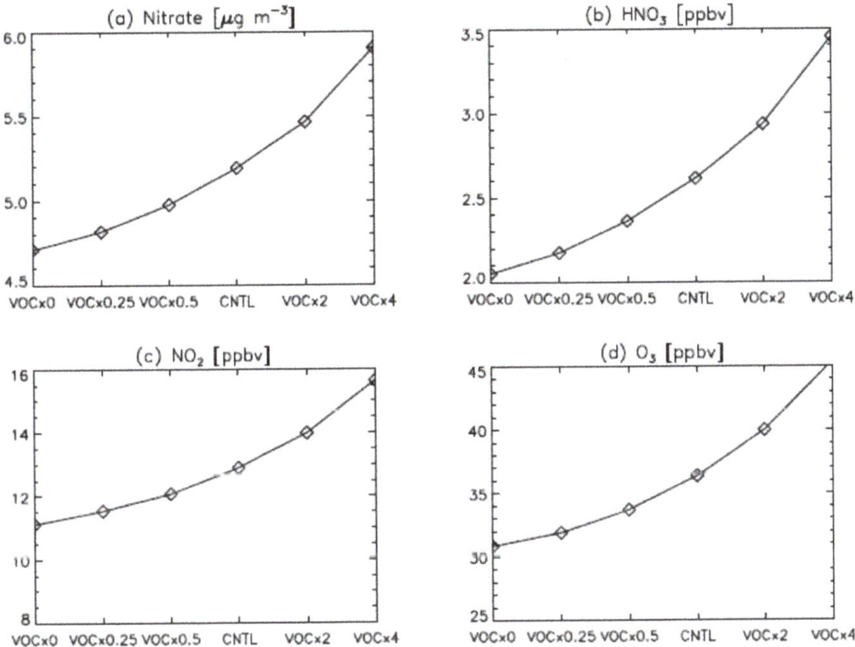

Figure 6. Average (**a**) nitrate, (**b**) HNO_3, (**c**) NO_2, and (**d**) O_3 concentrations in the street canyon (i.e., below 20 m) following changes in the emission rate of vehicular VOC.

Finally, we investigated the sensitivity of NH_3 emissions to nitrate production. Figure 7a shows the average concentration of nitrate aerosols in the street canyon following vehicular NH_3 emission changes. The concentration of nitrate aerosol was considerably influenced by the NH_3 emission changes; the nitrate concentration in the $NH_3 \times 4$ simulation was 42% higher than that in the standard simulation, and the nitrate concentration in $NH_3 \times 0.25$ was 85% of that in the standard simulation. These results indicate that the sensitivity of NH_3 emissions to nitration is much higher than that of NO_x and slightly higher than that of the VOC emissions. The HNO_3 concentrations are inversely proportional to NH_3 emissions, indicating that higher NH_3 emissions induce a higher conversion rate of HNO_3 to nitrate aerosol (Figure 7b). These results suggest that the production of ammonium nitrate is reduced by the low concentration of NH_3 under an NH_3-limited regime for nitrate production. Studies based on both modeling and observed campaigns have reported that nitrate formation occurs under an NH_3-limited regime in East Asian megacities, including SMR [21,60]. The nitrate aerosols in the $NH_3 \times 0$ simulation were 19% lower than those with the CNTL simulation, indicating that ammonium and NH_3 concentrations from the boundary also have an important role in nitrate formation in the urban street canyon. These results indicate that the control of NH_3 emissions might be the most effective way to degrade $PM_{2.5}$ problems where vehicular emissions are dominant in winter. The regulation of vehicle emissions is mostly focused on the control of NO_x emissions; considering the present findings, we should instead focus on controlling VOC and NH_3.

Though we used the coupled CFD–chemistry model to investigate the sensitivity of nitrate aerosols from vehicular emissions under complex geometry, our simulation still has limitations. Sea-salt aerosol significantly impacts the formation of nitrate aerosols via heterogeneous reactions when interacting with trace gases on the surface of sea-salt aerosol [61]. This process drives the efficient production of nitrates under an NH_3-limited environment. This model does not account for the effect of sea-salt aerosol on nitrate aerosol production. Therefore, the simulation might underestimate the nitrate formation of heterogeneous chemistry. Moreover, we only considered the effect of vehicular emissions; NH_3 and VOCs emissions from heating or biogenic emissions might affect the sensitivity of nitrate formation in the street canyon. The absence of these emissions in the model domain might create uncertainty in the nitrate aerosol calculation in this simulation.

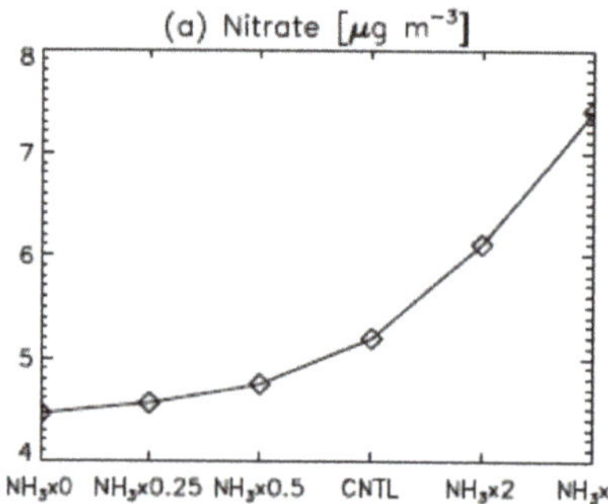
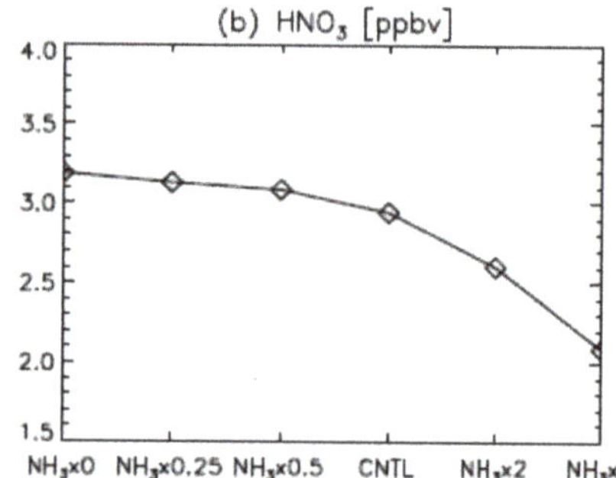

Figure 7. Average (**a**) nitrate and (**b**) HNO_3 concentrations in the street canyon (i.e., below 20 m) following a change in the emission rate of vehicular NH_3.

4. Model Sensitivity to Geometry and Speciation of VOC Emissions

We examined the sensitivity of the model to the canyon geometry by conducting sensitivity model simulations in which we changed the street canyon aspect ratios (the ratio of building height to street width) of the street canyon to 0.5 and 2.0. The conditions of the sensitivity simulations were identical to those of the CTNL simulation except that the height of the buildings, 10 m and 40 m, respectively, indicating canyon aspect ratios of 0.5 and 2.0 (Figure 8). We named the sensitivity simulations for different aspect ratios "species name" × "multiplying factor" _A "aspect ratio" (e.g., CNTL_A2.0 and NO_x × 2_A0.5). Figures 9 and 10 indicate the meridionally averaged NO_x, O_3, HNO_3, and nitrate aerosol concentrations in the CNTL_A0.5 and CNTL_A 2.0 simulations. The nitrate aerosol and their precursors showed similar distributions as those of the CNTL simulations despite the difference in their aspect ratios. The NO_x concentration was highest at the surface and is an order of magnitude higher than that outside the canyon, indicating the trapping of vehicular emissions due to the strong canyon vortex in the street canyon (Figures 9a and 10a). The spatial distribution of NO_x indicates that concentrated vehicular emissions drive the NO_x titration of the O_3 concentration at the surface under a low VOC emissions condition in both cases (Figures 9b and 10b). The HNO_3 concentrations also show similar distribution to those of the CTNL simulation, indicating the suppressing of the production of OH and HNO_3 (Figures 9c and 10c). Figures 9d and 10d show the nitrate concentrations in the street canyon for different canyon aspect ratios. The averaged nitrate concentrations in the street canyon (<10 m and <40 m, respectively) show 5.7 and 6.2 µg m^{-3} in the CNTL_A0.5 and CNTL_A 2.0 simulations, respectively, which are 9%, and 19% higher than those of the CTNL simulation. These differences are mainly due to the complex canyon vortex driving for the building geometry [27]. Despite the dispersion patterns differing according to canyon geometry, the mechanism of nitrate aerosol formation was consistent with that of CNTL, suggesting high nitrate formation from vehicular emissions in the street canyon.

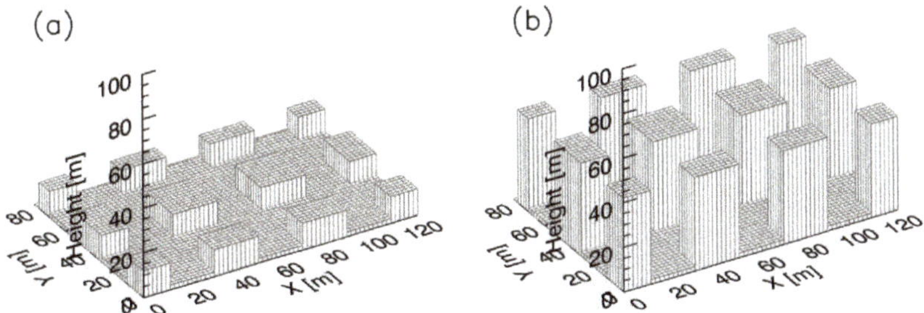

Figure 8. Schematic diagrams of the coupled sensitivity simulation domain for the canyon aspect ratios (**a**) 0.5 and (**b**) 2.0.

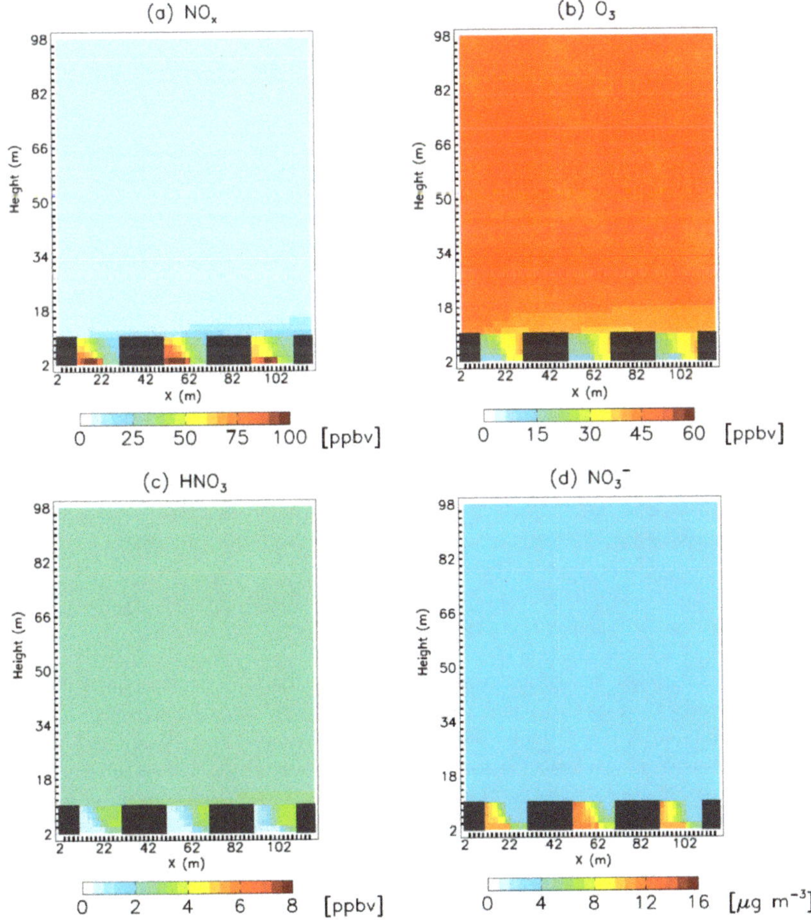

Figure 9. Distribution of the daily average concentrations of (**a**) NO$_x$, (**b**) HNO$_3$, (**c**) O$_3$ (ppbv), and (**d**) nitrate aerosol (μg m^{-3}) in the CNTL_A0.5 simulation.

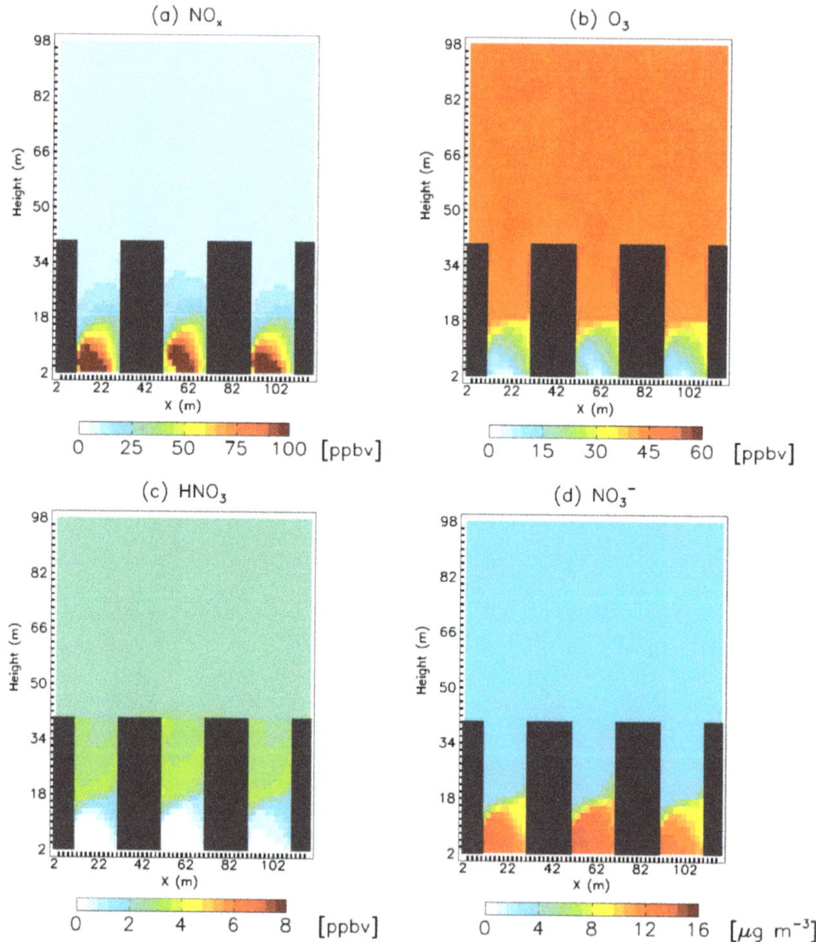

Figure 10. Distribution of the daily average concentrations of (**a**) NO_x, (**b**) HNO_3, (**c**) O_3 (ppbv), and (**d**) nitrate aerosol ($\mu g\ m^{-3}$) in the CNTL_A2.0 simulation.

We tested the NO_x, VOC, and NH_3 emission sensitivity to nitration formation for different canyon aspect ratios (0.5 and 2.0). Figures 11 and 12 show the averaged nitrate concentration in the street canyons (<10 m and <40 m, respectively) by following the vehicular NO_x, VOC, and NH_3 emission changes for the different canyon aspect ratios of 0.5 and 2.0, respectively. The sensitivity of the nitrate formation following NO_x, VOC, and NH_3 emission changes is also similar to those for the canyon aspect ratio of unity. The nitrate concentration changes show no clear relationship with the NO_x emission rate in either case, which is related to the nitrate precursor changes, particularly in the daytime as we mentioned (not shown). The maximum concentrations occurred in $NO_x \times 0.5\_A0.5$ and $NO_x \times 0.5\_A2.0$, which differed slightly with the simulations for the canyon aspect ratio of unity. Nevertheless, the difference in the nitrate concentrations between the simulations was only 2%. The sensitivity of nitrate formation following VOC and NH_3 emissions also follows consistent results with those for the aspect ratio of unity. Enhanced (reduced) VOC emissions drive an increase (decrease) in the nitrate aerosol concentration affecting the O_3 concentration under a VOC limited regime for O_3 production. The nitrate concentration in the street canyon is proportional to the vehicular NH_3 emission, indicating that NH_3 limits the condition of nitrate formation in both cases. These results indicate that the sensitivity of nitrate formation following emission changes in the street canyon is consistent, regardless of the building aspect ratio, due to concentrated emissions from vehicles and the canyon vortex.

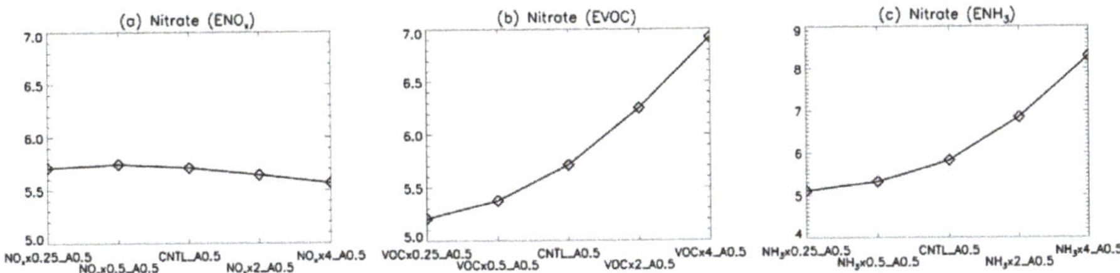

Figure 11. The average nitrate concentration in the street canyon following changes in the emission rate of vehicular (**a**) NO_x, (**b**) VOC, and (**c**) NH_3 for the 0.5 canyon aspect ratio; units are $\mu g\ m^{-3}$.

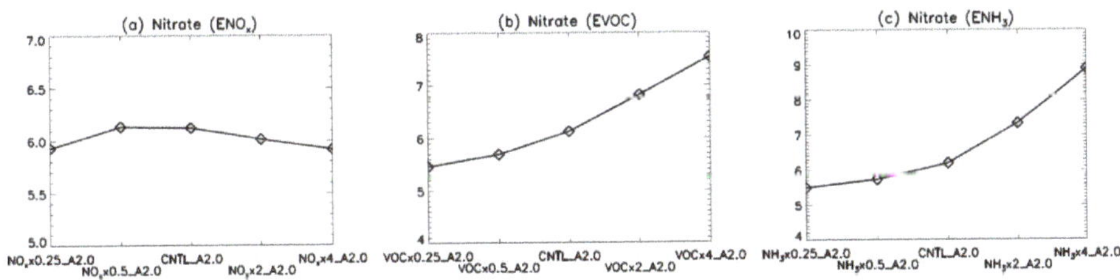

Figure 12. The same as Figure 11 but for the 2.0 canyon aspect ratio; units are $\mu g\ m^{-3}$.

The VOC speciation of vehicular emissions might change the model sensitivity on nitrate formation by affecting the ozone and OH production, considering all mechanisms are explained under a VOC limited regime for O_3 production. Therefore, we checked the sensitivity of VOC speciation of emissions on nitrate aerosol formation using the different VOC chemical speciation used by Kim et al. (2006) [62]. Table 5 summarizes the emission rates and the ratio of speciated VOC with the method of Kim et al. (2006). Figure 13 indicates the averaged nitrate concentration in the street canyon (i.e., below 20 m) by following vehicular NO_x, VOC, and NH_3 emission changes with the VOC speciation of Kim et al. (2006). Similar to other sensitivity simulations, NO_x emission changes did not affect the nitrate formation (due to the conflicting effects of NO_2 and O_3) even though we changed the VOC speciation (Figure 13a). The sensitivity of VOC concentration to nitrate formation also shows a similar relationship to that with EMEP/EEA speciation. However, the sensitivity was slightly lower than the nitrate concentration in CNTL, showing only a 6% difference between VOC × 0.25 and CNTL (Figure 13b). These results show that the reduction of vehicular VOC emission is a more effective way to regulate nitrate problems in urban streets than NO_x emissions under different VOC speciations.

Table 5. Emission rates per vehicle and ratios of speciated VOC obtained by following the method developed by Kim et al. (2006).

Tracer Name	Formula	Emission Rate [mg km^{-1}]	Ratio [%]
ALK4	$\geq C_4$ alkanes	6.03	40.2
ISOP	$CH_2 = C(CH_3)CH = CH_2$	-	-
ACET	$CH_3C(O)CH_3$	0.15	1.0
MEK	$RC(O)R$	0.0	0.0
ALD2	CH_3CHO	0.08	0.5
RCHO	CH_3CH_2CHO	0.08	0.5

Table 5. *Cont.*

Tracer Name	Formula	Emission Rate [mg km^{-1}]	Ratio [%]
MVK	CH$_2$ = CHC(=O)CH$_3$	-	-
MACR	CH$_2$ = C(CH$_3$)CHO	-	-
PRPE	$\geq$C$_3$ alkenes	2.10	14.0
C3H8	C$_3$H$_8$	0.15	1.0
CH2O	HCHO	0.17	1.1
C2H6	C$_2$H$_6$	0.27	1.8
Unspeciated	-	-	39.9

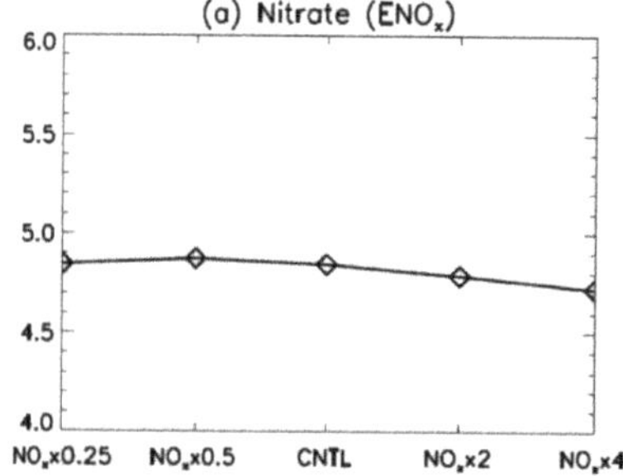

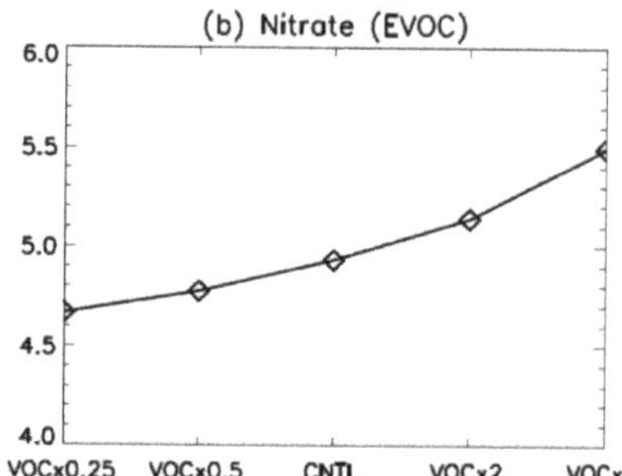

Figure 13. Average nitrate concentration in the street canyon following changes in the emission rate of vehicular (**a**) NO$_x$ and (**b**) VOC with the speciation method of VOC emission used by Kim et al. (2006); units are µg m^{-3}.

5. Conclusions

Nitrate contributions to PM$_{2.5}$ mass have increased in polluted urban areas, with an increasing number of severe haze events in East Asia. This study investigates the favorable conditions for the production of nitrate aerosols in urban streets using a coupled CFD–chemistry model. It was found that the nitrate concentrations in street canyons are higher than those outside the canyons due to the high HNO$_3$ concentrations from vehicular NO$_x$ in these canyons. However, the spatial pattern of nitrate aerosols in street canyons differs from that of HNO$_3$ due to NH$_3$, thus indicating that ammonium nitrate formation occurs under NH$_3$-limited conditions in street canyons.

Sensitivity simulations indicate that nitrate concentration does not show a clear relationship with the NO$_x$ emission rate, with nitrate changes of only 2% across among 16 time differences in NO$_x$ emissions. The HNO$_3$ concentration in street canyons changes in a similar manner to that of nitrate aerosols, indicating that the changes in nitrate aerosols in the sensitivity simulations are closely related to HNO$_3$ changes. An increase in the NO$_x$ emissions induces a decrease in O$_3$ and an increase in NO$_2$ under a VOC-limited regime for O$_3$ production. These changes in O$_3$ and NO$_2$ have a conflicting effect on the HNO$_3$ production in urban streets. Therefore, HNO$_3$ and nitrate aerosols have no linear relationship with NO$_x$ emissions and only undergo small changes. The sensitivity simulations were conducted by varying the vehicular VOC emissions to investigate their effect on nitrate production. The results show that nitrate concentrations are proportional to VOC emissions. Nitrate was decreased by 9% in the VOC × 0.25 simulation, indicating a relatively high sensitivity compared to that of NO$_x$. Decreased VOC emissions drive a decrease in both NO$_2$ and O$_3$ concentrations, creating unfavorable conditions for HNO$_3$ production, unlike the effect of changes in NO$_x$ emissions. The nitrate aerosol concentration is considerably influenced by NH$_3$ emissions, which show a higher sensitivity to nitrate production than do NO$_x$ and VOC emissions in urban streets. The nitrate concentration is proportional to NH$_3$ emissions with the additional production of ammonium nitrate under an NH$_3$-limited regime for nitrate production. This research implies that, where vehicular emissions are dominant in winter, the control of vehicular VOC and NH_3 emissions might be a more effective way to degrade $PM_{2.5}$ problems than controlling NO_x.

We checked the model sensitivity by changing the model's building geometry and VOC speciation. The sensitivity of nitrate formation by following emissions changes acts in a similar direction as CNTL simulation despite changing the building geometry and speciation of VOC emissions. The sensitivity simulations revealed that our results about the sensitivity of nitrate production to emission changes are robust.

Funding: This work was funded by the 2017 Research Fund of Myongji University.

Acknowledgments: This work was supported by the 2017 Research Fund of Myongji University.

Conflicts of Interest: The authors declare no conflict of interest.

References

1. Huang, R.-J.; Zhang, Y.; Bozzetti, C.; Ho, K.-F.; Cao, J.-J.; Han, Y.; Daellenbach, K.R.; Slowik, J.G.; Platt, S.M.; Canonaco, F. High secondary aerosol contribution to particulate pollution during haze events in China. *Nature* **2014**, *514*, 218. [PubMed]
2. Liu, X.; Sun, K.; Qu, Y.; Hu, M.; Sun, Y.; Zhang, F.; Zhang, Y. Secondary formation of sulfate and nitrate during a haze episode in megacity Beijing, China. *Aerosol Air Qual. Res.* **2015**, *15*, 2246–2257. [CrossRef]
3. Pan, Y.; Wang, Y.; Zhang, J.; Liu, Z.; Wang, L.; Tian, S.; Tang, G.; Gao, W.; Ji, D.; Song, T. Redefining the importance of nitrate during haze pollution to help optimize an emission control strategy. *Atmos. Environ.* **2016**, *141*, 197–202. [CrossRef]
4. Tan, J.; Duan, J.; He, K.; Ma, Y.; Duan, F.; Chen, Y.; Fu, J. Chemical characteristics of PM2.5 during a typical haze episode in Guangzhou. *J. Environ. Sci.* **2009**, *21*, 774–781.
5. Wang, H.-J.; Chen, H.-P. Understanding the recent trend of haze pollution in eastern China: Roles of climate change. *Atmos. Chem. Phys.* **2016**, *16*, 4205–4211. [CrossRef]
6. Han, S.H.; Kim, Y.P. Long-term Trends of the Concentrations of Mass and Chemical Composition in PM 2.5 over Seoul. *J. Korean Soc. Atmos. Environ.* **2015**, *31*, 143–156. [CrossRef]
7. Guo, H.; Ding, A.J.; So, K.L.; Ayoko, G.; Li, Y.S.; Hung, W.T. Receptor modeling of source apportionment of Hong Kong aerosols and the implication of urban and regional contribution. *Atmos. Environ.* **2009**, *43*, 1159–1169. [CrossRef]
8. Liu, T.; Wang, X.; Hu, Q.; Deng, W.; Zhang, Y.; Ding, X.; Fu, X.; Bernard, F.; Zhang, Z.; Lü, S. Formation of secondary aerosols from gasoline vehicle exhaust when mixing with SO_2. *Atmos. Chem. Phys.* **2016**, *16*, 675–689. [CrossRef]
9. Nowak, J.B.; Neuman, J.A.; Bahreini, R.; Middlebrook, A.M.; Holloway, J.S.; McKeen, S.A.; Parrish, D.D.; Ryerson, T.B.; Trainer, M. Ammonia sources in the California South Coast Air Basin and their impact on ammonium nitrate formation. *Geophys. Res. Lett.* **2012**, *39*. [CrossRef]
10. Durant, J.L.; Ash, C.A.; Wood, E.C.; Herndon, S.C.; Jayne, J.T.; Knighton, W.B.; Canagaratna, M.R.; Trull, J.B.; Brugge, D.; Zamore, W. Short-term variation in near-highway air pollutant gradients on a winter morning. *Atmos. Chem. Phys.* **2010**, *10*, 5599. [CrossRef] [PubMed]
11. Hagler, G.S.W.; Baldauf, R.W.; Thoma, E.D.; Long, T.R.; Snow, R.F.; Kinsey, J.S.; Oudejans, L.; Gullett, B.K. Ultrafine particles near a major roadway in Raleigh, North Carolina: Downwind attenuation and correlation with traffic-related pollutants. *Atmos. Environ.* **2009**, *43*, 1229–1234. [CrossRef]
12. Hu, S.; Fruin, S.; Kozawa, K.; Mara, S.; Paulson, S.E.; Winer, A.M. A wide area of air pollutant impact downwind of a freeway during pre-sunrise hours. *Atmos. Environ.* **2009**, *43*, 2541–2549. [CrossRef]
13. Xie, S.; Zhang, Y.; Qi, L.; Tang, X. Spatial distribution of traffic-related pollutant concentrations in street canyons. *Atmos. Environ.* **2003**, *37*, 3213–3224. [CrossRef]
14. Myhre, G.; Grini, A.; Metzger, S. Modelling of nitrate and ammonium-containing aerosols in presence of sea salt. *Atmos. Chem. Phys.* **2006**, *6*, 4809–4821. [CrossRef]

15. Riemer, N.; Vogel, H.; Vogel, B.; Schell, B.; Ackermann, I.; Kessler, C.; Hass, H. Impact of the heterogeneous hydrolysis of N2O5 on chemistry and nitrate aerosol formation in the lower troposphere under photosmog conditions. *J. Geophys. Res. Atmos.* **2003**, *108*.

16. Schaap, M.; Cuvelier, C.; Hendriks, C.; Bessagnet, B.; Baldasano, J.M.; Colette, A.; Thunis, P.; Karam, D.; Fagerli, H.; Graff, A. Performance of European chemistry transport models as function of horizontal resolution. *Atmos. Environ.* **2015**, *112*, 90–105. [CrossRef]

17. Tewari, M.; Kusaka, H.; Chen, F.; Coirier, W.J.; Kim, S.; Wyszogrodzki, A.A.; Warner, T.T. Impact of coupling a microscale computational fluid dynamics model with a mesoscale model on urban scale contaminant transport and dispersion. *Atmos. Res.* **2010**, *96*, 656–664. [CrossRef]

18. Wehner, M.F.; Reed, K.A.; Li, F.; Bacmeister, J.; Chen, C.-T.; Paciorek, C.; Gleckler, P.J.; Sperber, K.R.; Collins, W.D.; Gettelman, A. The effect of horizontal resolution on simulation quality in the Community Atmospheric Model, CAM5.1. *J. Adv. Model. Earth Syst.* **2014**, *6*, 980–997. [CrossRef]

19. Prasad, R.; Bella, V.R. A review on diesel soot emission, its effect and control. *Bull. Chem. React. Eng. Catal.* **2011**, *5*, 69–86. [CrossRef]

20. Zhang, Q.; He, K.; Huo, H. Policy: Cleaning China's air. *Nature* **2012**, *484*, 161.

21. Link, M.F.; Kim, J.; Park, G.; Lee, T.; Park, T.; Babar, Z.B.; Sung, K.; Kim, P.; Kang, S.; Kim, J.S. Elevated production of NH4NO3 from the photochemical processing of vehicle exhaust: Implications for air quality in the Seoul Metropolitan Region. *Atmos. Environ.* **2017**, *156*, 95–101. [CrossRef]

22. Pathak, R.K.; Wu, W.S.; Wang, T. Summertime PM 2.5 ionic species in four major cities of China: Nitrate formation in an ammonia-deficient atmosphere. *Atmos. Chem. Phys.* **2009**, *9*, 1711–1722. [CrossRef]

23. Wen, L.; Chen, J.; Yang, L.; Wang, X.; Xu, C.; Sui, X.; Yao, L.; Zhu, Y.; Zhang, J.; Zhu, T. Enhanced formation of fine particulate nitrate at a rural site on the North China Plain in summer: The important roles of ammonia and ozone. *Atmos. Environ.* **2015**, *101*, 294–302. [CrossRef]

24. Pun, B.K.; Seigneur, C. Sensitivity of particulate matter nitrate formation to precursor emissions in the California San Joaquin Valley. *Environ. Sci. Technol.* **2001**, *35*, 2979–2987. [CrossRef]

25. Kim, M.J.; Park, R.J.; Kim, J.-J. Urban air quality modeling with full O$_3$–NO$_x$–VOC chemistry: Implications for O$_3$ and PM air quality in a street canyon. *Atmos. Environ.* **2012**, *47*, 330–340. [CrossRef]

26. Kim, J.-J.; Baik, J.-J. A numerical study of the effects of ambient wind direction on flow and dispersion in urban street canyons using the RNG k–ε turbulence model. *Atmos. Environ.* **2004**, *38*, 3039–3048. [CrossRef]

27. Park, S.-J.; Kim, J.-J.; Kim, M.J.; Park, R.J.; Cheong, H.-B. Characteristics of flow and reactive pollutant dispersion in urban street canyons. *Atmos. Environ.* **2015**, *108*, 20–31. [CrossRef]

28. Kim, M.J.; Park, R.J.; Kim, J.-J.; Park, S.H.; Chang, L.-S.; Lee, D.-G.; Choi, J.-Y. A computational fluid dynamics simulation of reactive fine particulate matter in street canyons. *Atmos. Environ.* **2019**, *209*, 54–66. [CrossRef]

29. Bey, I.; Jacob, D.J.; Yantosca, R.M.; Logan, J.A.; Field, B.D.; Fiore, A.M.; Li, Q.; Liu, H.Y.; Mickley, L.J.; Schultz, M.G. Global modeling of tropospheric chemistry with assimilated meteorology: Model description and evaluation. *J. Geophys. Res. Atmos.* **2001**, *106*, 23073–23095. [CrossRef]

30. Wang, Y.; Zhang, Q.; Jiang, J.; Zhou, W.; Wang, B.; He, K.; Duan, F.; Zhang, Q.; Philip, S.; Xie, Y. Enhanced sulfate formation during China's severe winter haze episode in January 2013 missing from current models. *J. Geophys. Res. Atmos.* **2014**, *119*, 10–425. [CrossRef]

31. Yang, Y.; Liao, H.; Lou, S. Increase in winter haze over eastern China in recent decades: Roles of variations in meteorological parameters and anthropogenic emissions. *J. Geophys. Res. Atmos.* **2016**, *121*, 13050–13065. [CrossRef]

32. Neu, J.L.; Prather, M.J.; Penner, J.E. Global atmospheric chemistry: Integrating over fractional cloud cover. *J. Geophys. Res. Atmos.* **2007**, *112*. [CrossRef]

33. Wild, O.; Zhu, X.; Prather, M.J. Fast-J: Accurate simulation of in-and below-cloud photolysis in tropospheric chemical models. *J. Atmos. Chem.* **2000**, *37*, 245–282. [CrossRef]

34. Park, R.J.; Jacob, D.J.; Field, B.D.; Yantosca, R.M.; Chin, M. Natural and transboundary pollution influences on sulfate-nitrate-ammonium aerosols in the United States: Implications for policy. *J. Geophys. Res. Atmos.* **2004**, *109*. [CrossRef]

35. Park, R.J.; Jacob, D.J.; Chin, M.; Martin, R.V. Sources of carbonaceous aerosols over the United States and implications for natural visibility. *J. Geophys. Res. Atmos.* **2003**, *108*. [CrossRef]

36. Fountoukis, C.; Nenes, A. ISORROPIA II: A computationally efficient thermodynamic equilibrium model for K^+–Ca^{2+}–Mg^{2+}–NH^{4+}–Na^+–SO_4^2–NO_3–Cl–H_2O aerosols. *Atmos. Chem. Phys.* **2007**, *7*, 4639–4659. [CrossRef]

37. Nenes, A.; Pandis, S.N.; Pilinis, C. ISORROPIA: A new thermodynamic equilibrium model for multiphase multicomponent inorganic aerosols. *Aquat. Geochem.* **1998**, *4*, 123–152. [CrossRef]

38. Jacob, D.J. Heterogeneous chemistry and tropospheric ozone. *Atmos. Environ.* **2000**, *34*, 2131–2159. [CrossRef]

39. Zhang, L.; Jacob, D.J.; Knipping, E.M.; Kumar, N.; Munger, J.W.; Carouge, C.C.; Van Donkelaar, A.; Wang, Y.X.; Chen, D. Nitrogen deposition to the United States: Distribution, sources, and processes. *Atmos. Chem. Phys.* **2012**, *12*, 4539–4554. [CrossRef]

40. Walker, J.M.; Philip, S.; Martin, R.V.; Seinfeld, J.H. Simulation of nitrate, sulfate, and ammonium aerosols over the United States. *Atmos. Chem. Phys.* **2012**, *12*, 11213–11227. [CrossRef]

41. Cooke, W.F.; Liousse, C.; Cachier, H.; Feichter, J. Construction of a 1 × 1 fossil fuel emission data set for carbonaceous aerosol and implementation and radiative impact in the ECHAM4 model. *J. Geophys. Res. Atmos.* **1999**, *104*, 22137–22162. [CrossRef]

42. Wesely, M.L. Parameterization of surface resistances to gaseous dry deposition in regional-scale numerical models. *Atmos. Environ. (1967)* **1989**, *23*, 1293–1304. [CrossRef]

43. Lee, D.-G.; Lee, Y.-M.; Jang, K.-W.; Yoo, C.; Kang, K.-H.; Lee, J.-H.; Jung, S.-W.; Park, J.-M.; Lee, S.-B.; Han, J.-S. Korean national emissions inventory system and 2007 air pollutant emissions. *Asian J. Atmos. Environ.* **2011**, *5*, 278–291. [CrossRef]

44. Carslaw, D.C.; Beevers, S.D. Estimations of road vehicle primary NO2 exhaust emission fractions using monitoring data in London. *Atmos. Environ.* **2005**, *39*, 167–177. [CrossRef]

45. Amon, B.; Hutchings, N.; Dämmgen, U.; Webb, J. *EMEP/EEA Air Pollutant Emission Inventory Guidebook—2016*; European Environment Agency: Copenhagen, Denmark, 2016.

46. Heo, J.-B.; Hopke, P.K.; Yi, S.-M. Source apportionment of PM 2.5 in Seoul, Korea. *Atmos. Chem. Phys.* **2009**, *9*, 4957–4971. [CrossRef]

47. Shin, H.J.; Roh, S.A.; Kim, J.C.; Lee, S.J.; Kim, Y.P. Temporal variation of volatile organic compounds and their major emission sources in Seoul, Korea. *Environ. Sci. Pollut. Res.* **2013**, *20*, 8717–8728. [CrossRef]

48. Shon, Z.-H.; Kim, K.-H.; Song, S.-K.; Jung, K.; Kim, N.-J.; Lee, J.-B. Relationship between water-soluble ions in $PM_{2.5}$ and their precursor gases in Seoul megacity. *Atmos. Environ.* **2012**, *59*, 540–550. [CrossRef]

49. Kim, Y.; Guldmann, J.-M. Impact of traffic flows and wind directions on air pollution concentrations in Seoul, Korea. *Atmos. Environ.* **2011**, *45*, 2803–2810. [CrossRef]

50. Inness, A.; Baier, F.; Benedetti, A.; Bouarar, I.; Chabrillat, S.; Clark, H.; Clerbaux, C.; Coheur, P.; Engelen, R.J.; Errera, Q. The MACC reanalysis: An 8 yr data set of atmospheric composition. *Atmos. Chem. Phys.* **2013**, *13*, 4073–4109. [CrossRef]

51. Huang, M.; Carmichael, G.R.; Pierce, R.B.; Jo, D.S.; Park, R.J.; Flemming, J.; Emmons, L.K.; Bowman, K.W.; Henze, D.K.; Davila, Y. Impact of intercontinental pollution transport on North American ozone air pollution: An HTAP phase 2 multi-model study. *Atmos. Chem. Phys.* **2017**, *17*, 5721–5750. [CrossRef]

52. Kim, J.-J.; Baik, J.-J. Effects of street-bottom and building-roof heating on flow in three-dimensional street canyons. *Adv. Atmos. Sci.* **2010**, *27*, 513–527. [CrossRef]

53. Baker, J.; Walker, H.L.; Cai, X. A study of the dispersion and transport of reactive pollutants in and above street canyons—A large eddy simulation. *Atmos. Environ.* **2004**, *38*, 6883–6892. [CrossRef]

54. Yubero, E.; Galindo, N.; Nicolás, J.F.; Crespo, J.; Calzolai, G.; Lucarelli, F. Temporal variations of PM1 major components in an urban street canyon. *Environ. Sci. Pollut. Res.* **2015**, *22*, 13328–13335. [CrossRef] [PubMed]

55. Lu, Z.; Streets, D.G.; Zhang, Q.; Wang, S.; Carmichael, G.R.; Cheng, Y.F.; Wei, C.; Chin, M.; Diehl, T.; Tan, Q. Sulfur dioxide emissions in China and sulfur trends in East Asia since 2000. *Atmos. Chem. Phys.* **2010**, *10*, 6311–6331. [CrossRef]

56. WHO. Review of Evidence on Health Aspects of Air Pollution–REVIHAAP Project. Available online: https://scholar.google.com/scholar_lookup?title=Review%20of%20evidence%20on%20health%20aspects%20of%20air%20pollution%20%E2%80%94%20REVIHAAP%20project%20technical%20report&publication_year=2013&author=WHOhttps://scholar.google.co.kr (accessed on 2 February 2019).

57. Nguyen, K.; Dabdub, D. NOx and VOC control and its effects on the formation of aerosols. *Aerosol Sci. Technol.* **2002**, *36*, 560–572. [CrossRef]

58. Jin, X.; Fiore, A.M.; Murray, L.T.; Valin, L.C.; Lamsal, L.N.; Duncan, B.; Folkert Boersma, K.; De Smedt, I.; Abad, G.G.; Chance, K. Evaluating a Space-Based Indicator of Surface Ozone-NOx-VOC Sensitivity Over Midlatitude Source Regions and Application to Decadal Trends. *J. Geophys. Res. Atmos.* **2017**, *122*.

59. Li, K.; Jacob, D.J.; Liao, H.; Shen, L.; Zhang, Q.; Bates, K.H. Anthropogenic drivers of 2013–2017 trends in summer surface ozone in China. *Proc. Natl. Acad. Sci. USA* **2019**, *116*, 422–427. [CrossRef]

60. Chang, W.; Zhan, J.; Zhang, Y.; Li, Z.; Xing, J.; Li, J. Emission-driven changes in anthropogenic aerosol concentrations in China during 1970–2010 and its implications for PM2.5 control policy. *Atmos. Res.* **2018**, *212*, 106–119. [CrossRef]

61. Chen, Y.; Cheng, Y.; Ma, N.; Wolke, R.; Nordmann, S.; Schüttauf, S.; Ran, L.; Wehner, B.; Birmili, W.; Gon, H.A. Sea salt emission, transport and influence on size-segregated nitrate simulation: A case study in northwestern Europe by WRF-Chem. *Atmos. Chem. Phys.* **2016**, *16*, 12081–12097. [CrossRef]

62. Kim, M.-S.; Kim, J.H.; Park, H.-S.; Sun, Y.S.; Kim, H.-S.; Choi, K.H.; Yi, J. Emission inventory of VOCs from mobile sources in a metropolitan region. *Korean J. Chem. Eng.* **2006**, *23*, 919–924. [CrossRef]

MDPI
St. Alban-Anlage 66
4052 Basel
Switzerland
Tel. +41 61 683 77 34
Fax +41 61 302 89 18
www.mdpi.com

Atmosphere Editorial Office
E-mail: atmosphere@mdpi.com
www.mdpi.com/journal/atmosphere